Modeling nutrient and moisture cycling in tropical forests

CIP-DATA KONINKLIJKE BIBLIOTHEEK, DEN HAAG

Modeling

Modeling nutrient and moisture cycling in tropical forests // I.G.A.M. Noij ... [et al.]. - Wageningen: The Tropenbos Foundation. - Ill. - (Tropenbos Series)
With index, ref.
ISBN 90-5113-016-3
Subject headings: nutrient cycling / nutrient concentrations / tropical rain forests.

Cover design	:	Diamond Communications
Printed by	:	Veenman Drukkers, Wageningen
Photography	:	Mountain forest, Bukit Raya Nature Reserve, Central Kalimantan Province, Indonesia (photo by H. van Reuler)

MODELING NUTRIENT AND MOISTURE CYCLING IN TROPICAL FORESTS

I.G.A.M. Noij, B.H. Janssen,
L.G. Wesselink and J.J.M. van Grinsven

The Tropenbos Foundation
Wageningen, The Netherlands
1993

TROPENBOS SERIES

The Tropenbos Series presents the results of studies and research activities related to the conservation and wise utilization of forest lands in the humid tropics. The series continues and integrates the former Tropenbos Scientific and Technical Series. The studies published in this series have been carried out within the international Tropenbos programme. Occasionally, this series may present the results of other studies which contribute to the objectives of the Tropenbos programme.

TROPENBOS

CONTENTS

PART III MODEL DESCRIPTION AND SENSITIVITY ANALYSIS

PREFACE

The basic philosophy of the Tropenbos Programme involves an integrated methodical approach comprising studies on the earth's physical and biotic aspects as well as social studies in humid tropical forests. To ensure comparability of the results of research at the various locations and also to facilitate transfer and extrapolation of these data, Tropenbos promotes the use of common research methodologies.
In order to establish guidelines for a common methodology for nutrient and moisture cycling in tropical forests, the *Working Group for a Common Methodology on Nutrient and Moisture Cycling* was initiated in January 1987. This working group consisted of soil scientists, geologists, vegetation experts/ecologists and foresters of the University of Amsterdam (Laboratory of Physical Geography and Soil Science), the Wageningen Agricultural University (Department of Forestry and Forest Ecology; Department of Soil Science and Geology; Department of Soil Science and Plant Nutrition), the University of Groningen, (Department of Physical Geography and Soil Science), the International Soil Reference and Information Centre (ISRIC) at Wageningen, the Institute for Soil Fertility, (IB) at Haren, the Royal Tropical Institute at Amsterdam, the Netherlands Soil Survey Institute (STIBOKA) at Wageningen, Tropenbos Wageningen, and UNESCO at Paris.

Meetings of this working group brought together the activities of various professionally-related but hitherto independently operating institutions. In the context of the working group's aim to standardize methods and outputs of research, the meetings generated discussions as to the most desirable, but still practical, approach to study nutrient and moisture cycling in humid tropical forest areas.
In these meetings the following subjects were discussed:
- observation density in space and time and standardization of sampling;
- type and procedure of laboratory analyses;
- modeling of nutrient and moisture cycling.

In an early stage, the working group recognized the usefulness of models. In order to expedite the development of a model, Tropenbos made funds available for the Wageningen Agricultural University to contract scientists for that purpose.

This report forms the output of a project entitled "Understanding Nutrient and Moisture Cycling in Humid Tropical Forest Land (Phase I)".

The objective of the project was to develop a simulation model that

- gives a quantitative description of moisture and nutrient cycling in Tropical Rain Forest and if possible in derived ecosystems, like those under shifting cultivation, tree-crops, permanent cultivation of annual crops;
- calculates growth of vegetation (primary forest, secondary vegetation, tree-crops, annual crops) as output of nutrient and moisture cycling.

This project can be considered as the follow-up of the work by the mentioned Working Group on Nutrients and Moisture Cycling (Van Breemen, 1988), and of a preceding exercise in nutrient modeling at the Department of Soil Science and Plant Nutrition of the Agricultural University of Wageningen (Noij, 1988; Noij et al., 1988). At the same department the present model was developed by ir. I.G.A.M. Noij (soil fertility, crop growth modeling), dr.ir. B.H. Janssen (fertility of tropical soils, modeling of (soil) organic-matter mineralization), ir. L.G. Wesselink (modeling of ecopedological processes) and dr.ir. J.J.M. van Grinsven (modeling of (soil) chemical and hydrological processes). Prof.dr.ir. N. van Breemen (Wageningen Agricultural University) and prof.dr. J.M. Verstraten (University of Amsterdam) acted as advisors on ad-hoc basis.

It was felt practical to subdivide the reporting on the project into three parts:

- Part I gives the headlines of the model, and indicates its strong and weak points and possible uses. It is meant for interested scientists and officers who themselves are not specialized in modeling.
- Part II consists of a literature study on nutrient relationships in especially the vegetation compartment of the rain forest ecosystem. This study was needed to find data and elaborate them as to make them suited for incorporation in the model.
- Part III describes the model in its very details. It forms the heart of the matter and can be regarded as a background document for those who want to use the model, understand the code, modify the code, or use model formulations for other programs.

The report is the result of a joint action of the four authors. Janssen acted as editor and is the main author of Part I, Noij is the main author of Part II, and Wesselink and Van Grinsven are the main authors of Part III and of the computer programme.

The authors are indebted to dr. W.G. Sombroek, who in 1986 as member of the Tropenbos Programme Commission, initiated both our 'Working Group on Nutrient and Moisture Cycling' and the sister 'Working Group for a Common Methodology on Land Inventory and Forest Land Evaluation'. Thanks are also due to all working group participants. Dr. W.B.J. Jonkers, Tropenbos, and Mr. V.G. Jetten, University of Utrecht, Department of Earth Sciences contributed substantially to the readability of the text, by their much appreciated critical comments on the manuscript. Last but not least thanks go to Mrs. M. Slootman-Vermeer for her patience when typing the manuscript.

More information on the scientific background of the model can be obtained from or via the project coordinator Dr. B.H. Janssen, Department of Soil Science and Plant Nutrition, Wageningen Agricultural University, P.O. Box 8005, 6700 EC Wageningen, The Netherlands.
(The cited literature can be found in the References list of Part I).

PART I

SHORT DESCRIPTION AND EVALUATION OF THE MODEL

CONTENTS

INTRODUCTION

Tropical forest research has often been scattered, mono-disciplinary and of short duration, without effective application in the area itself, or without transfer of results of the regions. This applies not only to many research programmes of institutions in the tropics, where it is caused by repeated shortage of funds, lack of trained manpower, etc. It also, and maybe even more so, prevails at European and North American research centres and international development agencies, which often focus on scientific publications, or lack contact with local institutions dealing with land use planning. As a result, the information gathered is not comparable, or it does not get to the right people.

Tropenbos intends to contribute to a more systematic and more interdisciplinary research approach which should result in sound land-use planning aimed at using tropical forests and forest lands on a sustainable basis, while safeguarding unique ecological values and macro-environmental functions of the regions concerned.
Transfer of information and experience on the use of tropical forests will be facilitated, within and between countries, by developing and promoting a *common overall approach*, as well as *common methodologies per discipline*. Sombroek (1986) emphasized the need for common methodologies, and launched the 'five-step' approach, discussed below.
The Programme can also lead to the strengthening of research centres and development in the tropical countries concerned, through assistance and participation in an international network.
On the one hand, there is a suggested framework of consecutive steps to be taken at the level of locational research while, on the other hand there are aspects inherent to the networking concept.
The following 'five-step' approach of research and planning has been proposed for all Tropenbos locations.

Step 1
Identification of a representative area and broad inventory of its resources.

Step 2
Aggregation of the inventory data and their qualitative evaluation, delineation of major land units and selection of sites for further detailed studies.

Step 3
Detailed quantitative investigation of ecological characteristics, potentials and regulatory processes at the selected sites.
This step would consist of subjects such as forest structure, animal behaviour, growth and yield, natural regeneration, entomology, species diversity, biomass and organic matter, nutrient and moisture cycles, and forms of degradation.

Step 4
Quantitative studies on human activities and needs, in relation to alternative forms of land use and aggregation of data into a characterization of land use systems.

Step 5
Integration of all information into a quantitative multi-purpose land evaluation, and the design of suitable options for the management of the main forest land units.

The present study deals mainly with Step 3 of the 'five-step' approach, and tries to fulfil the networking requirements of the common methodologies mentioned above. It is not a description of sampling and analytical procedures. For such information Tropenbos makes use of the handbook by Anderson and Ingram (1989).

Any management strategy for sustained land use of former or present tropical forest areas must be based on an understanding of the interactions between climate, vegetation and soil. To acquire such an understanding, the researcher may greatly benefit from modeling approaches, as was outlined by the Tropenbos Working Group on Nutrient and Moisture Cycling (Van Breemen, 1988).

When designing models, it clearly comes to light what information is still missing. On the other hand, in models maximum advantage is taken of the information available. Moreover, by sensitivity analysis models can indicate which parameters and processes are of vital importance, and thus worthwhile to study. In this way models assist in strategic planning of research activities. One step further is the use of models for (management) scenario analyses. Some examples are calculation of the effects of different timber felling regimes, the effects of different lengths of crop cultivation and fallow periods in shifting-cultivation systems, and the effects of burning. The ultimate goal of our efforts in modeling nutrient and moisture cycling is to make such long-term analyses possible.

The model resulting from the present study originated from the model NUTCYC, described in detail by Noij (1988) and summarized by Noij et al. (1988). In NUTCYC, the availability of phosphorus is taken as the driving force for plant growth. This choice was based on results obtained in the Taï region in Côte d' Ivoire (Van Reuler & Janssen, 1988; 1989). The literature review in Part II, Section 1.2, of the present report confirms that phosphorus often is a major growth-limiting factor. This is certainly the case in Sarawak, as shown by Staritsky (1988), who applied NUTCYC to the data gathered in a monitoring project on nutrient cycling by Andriesse (1987). It goes too far, however, to proclaim phosphorus-limited growth to universal rule for tropical forests. On sandy soils, cations like calcium may be growth limiting as well (Part II, Section 1.2 of this report), whereas in montane forests nitrogen may be critical (Jordan, 1985). In humid tropics, scarcity of nutrients generally is the main growth-regulating factor. But even in areas with abundant rainfall, plant growth might be reduced by moisture deficiency during dry spells.

These facts made it necessary to modify NUTCYC in such a way that shortage of other nutrients than phosphorus, periodical droughts and waterlogging can be taken into account. Another important advantage of the present model is that it calculates leaf-fall rate and nutrient concentrations in plant components in a dynamic way, while in NUTCYC these factors were fixed input data.

Introducing the moisture regime in the model meant that the time step of one year as used in NUTCYC could not be maintained. In the present model the time step may vary from two weeks to one year, according to the requirements ordained by the prevailing environmental conditions and the availability of data. The model calculates for each time step which factor is most limiting, and is therefore the driving force for plant growth during that time step. It is important to note that solar radiation is not dealt with in our model, in contrast to the simulation model by Mohren (1987) for temperate regions. Approximate calculations have indicated that in the tropics solar radiation may under certain conditions be more limiting than moisture or nutrients. Such conditions occur only a few days per year and are therefore of minor importance. Inclusion of potential dry-matter production would greatly extend and complicate the model, making it impossible to finish the project within the time schedule foreseen.

Models are simplifications of reality. This is certainly true when complicated systems like tropical forests are the subject of modeling. As main output the present model delivers the biomass of leaves, stems and roots for a forest vegetation as a whole at any time, but it leaves out the distribution of biomass among species. Biomass is calculated for a particular site with a given set of soil and climate characteristics (point analysis). Not yet considered are multi-dimensional systems such as topo-sequences and watersheds. Restrictions were also necessary with regard to the number of soil layers and nutritive elements. In Section 4 it is indicated how the model can be extented to improve its applicability, and what field measurements should be done for both testing and feeding the model.

The model is called DYNAMITE, which stands for DYnamics of Nutrients And Moisture In Tropical Ecosystems.

2 MODEL STRUCTURE

2.1 State variables

In the model, the forest ecosystem is subdivided into three main compartments (Fig. I.1):

- vegetation, consisting of leaves, wood, coarse roots, fine roots;
- forest floor, subdivided into leaf litter and wood litter;
- soil, consisting of one or more soil layers.

For the calculation of water flows, usually the soil profile is subdivided into more layers than for the calculation of nutrient flows. In the present version of the model, these numbers are two and one, respectively, but this can be modified. The soil layer distinguished for the calculation of nutrient flows contains organic and inorganic components. The organic component consists of fine-root debris, coarse-root debris, organic labile pool, organic moderately labile pool, organic stable pool. The inorganic component consists of inorganic labile pool, inorganic stable pool, inorganic inert pool. Each pool may contain one or more elements. In the present version of the model these are N, P, K and C. The model also distinguishes a soil solution, which for the upper soil layer must be interpreted as the total amount of water present in the forest floor and the upper soil layer. Plants can take up nutrients from the soil solution only.

2.2 Fluxes

2.2.1 Water

Water enters the ecosystem via rainfall, and leaves it via evaporation, transpiration and percolation. Part of the precipitation is intercepted by the vegetation and evaporates. The remaining part (throughfall) reaches the soil and infiltrates into the upper soil layer.

From the water that has entered the upper soil layer part is used for direct evaporation and another part for transpiration by plants. What remains is stored in the soil layer. If the water content exceeds field capacity, percolation to the second layer takes place. If the infiltration into a soil layer exceeds the maximum storage, percolation and transpiration, back-flow of water to higher soil layers is calculated. Back-flow from the first layer is calculated as surface storage. This water can enter the soil during the next time step. Thus it is assumed that no runoff occurs. If the water content becomes less than field capacity, capillary rise is calculated. Distribution of uptake of moisture from the various layers is fixed.

2.2.2 Nutrients

The nutrients nitrogen, phosphorus and potassium are considered. Besides, carbon is taken into account as the main constituent of living and dead organic components. In the model, nutrients enter the ecosystem via dry (phosphorus and potassium) and wet (nitrogen and potassium) deposition. Carbon enters the system via the process of photosynthesis. This process is not explicitly simulated. The amount of carbon follows from growth vegetation (Section 2.2.2.5). Input of nitrogen by microbiological fixation of N_2 has not yet explicitly been included in the model. Time was not available to model this process.

Nutrients leave the ecosystem by leaching and erosion, while carbon disappears as CO_2 due to dissimilation of dead organic components (Section 2.2.2.2). Losses of CO_2 by respiration and gaseous losses of nitrogen by denitrification and ammonia volatilization are not explicitly dealt with.

Nutrient flows to and from the various pools within the ecosystem result from and result in transformations of these pools. To keep the entire complex of nutrient flows within the ecosystem surveyable, the following distinction is made:
- transformations of the inorganic pools;
- transformations of the organic pools;
- nutrient flows to and from the solution;
- uptake and distribution of nutrients by and in the vegetation.

Nutrient flows and vegetation growth are discussed in the following subsections.

2.2.2.1 Transformations of inorganic pools

Phosphorus
The inorganic pools for phosphorus are the inert, stable and labile pools (Fig. I.1). It is assumed that the inorganic inert pool does not weather. The only of this pool are input by dry deposition and loss by erosion.

The inorganic stable pool increases by dry deposition and by transformation of labile into stable phosphorus, and decreases by erosion and by transformation of stable into labile phosphorus. There is no direct flow from inorganic stable phosphorus to the solution.

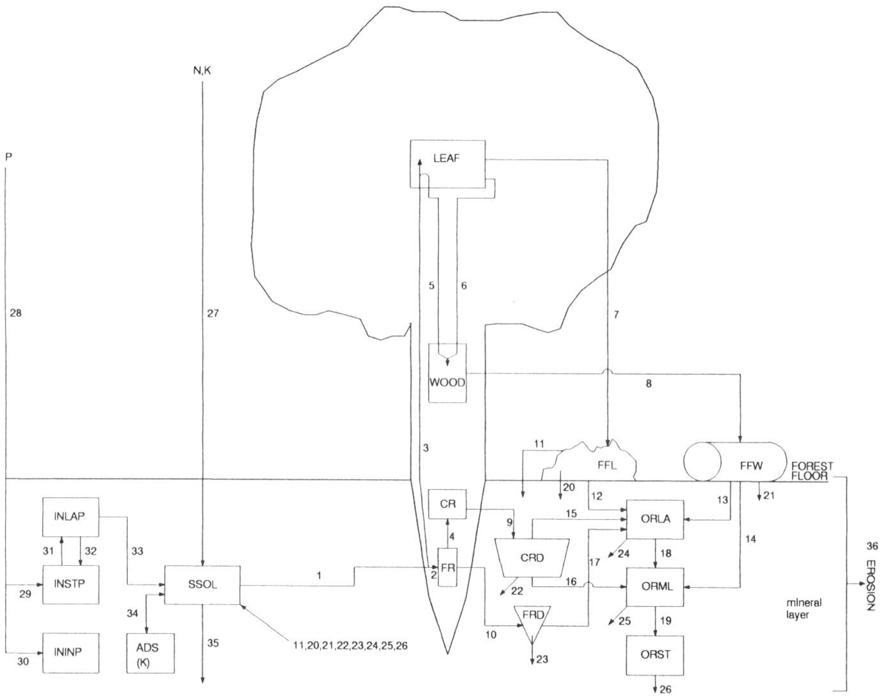

N,K
P
LEAF
5
6
28
27
7
WOOD
8
3
11
FFL
FFW
FOREST FLOOR
20
12
13
21
CR
15
ORLA
INLAP
9
31
32
33
4
CRD
14
36
EROSION
24
18
17
INSTP
SSOL
1
2
FR
16
ORML
29
22
mineral layer
34
FRD
25
19
ININP
ADS (K)
35
11,20,21,22,23,24,25,26
10
ORST
30
23
26

Fig. I.1 Schematic representation of the processes incorporated in the model DYNAMITE. The numbers refer to:

1	plant nutrient uptake from soil solution (SSOL)
2	nutrient uptake in fine roots (FR)
3	nutrient transport from fine roots to leaves (LEAF)
4	nutrient transition from fine to coarse roots (CR)
5	nutrient overflow from leaves to wood
6	nutrient retranslocation from leaves to wood
7	leaf fall
8	wood fall
9	coarse-root dying
10	fine-root sloughing
11	K leaching from forest-floor leaves (FFL)
12	nutrient transfer from forest-floor leaves to organic labile pool (ORLA)
13	nutrient transfer from forest-floor wood (FFW) to organic labile pool
14	nutrient transfer from forest-floor wood to organic moderately labile pool (ORML)
15	nutrient transfer from coarse-root debris (CRD) to organic labile pool
16	nutrient transfer from coarse-root debris to organic moderately labile pool
17	nutrient transfer from fine-root debris (FRD) to organic labile pool
18	nutrient transfer from organic labile to organic moderately labile pool
19	nutrient transfer from organic moderately labile to organic stable pool (ORST)
20	mineralization of forest floor leaves
21	mineralization of forest floor wood
22	mineralization of coarse-root debris
23	mineralization of fine-root debris
24	mineralization of organic labile pool
25	mineralization of organic moderately labile pool
26	mineralization of organic stable pool
27	wet deposition of N and K
28	total dry deposition of P
29	dry deposition of P to inorganic stable phosphorus (INSTP)
30	dry deposition of P to inorganic inert phosphorus (ININP)
31	transfer of P from inorganic stable to inorganic labile phosphorus (INLAP)
32	transfer of P from inorganic labile to inorganic stable phosphorus
33	transfer of P from inorganic labile pool to soil solution
34	desorption and adsorption of K
35	leaching of nutrients from soil solution (SSOL)
36	erosion of organic and inorganic nutrient pools

The inorganic labile pool increases by transformation of inorganic stable into inorganic labile phosphorus, and decreases by erosion, transformation of inorganic labile into inorganic stable phosphorus, and by dissolution.

Potassium

Two pools are distinghuished, namely exchangeable potassium (labile) and potassium containing minerals (stable). The labile pool increases by adsorption of potassium from the solution, and decreases by erosion and by desorption of adsorbed potassium. The stable pool decreases by weathering and increases by dry deposition. These processes are not explicitly modeled. Weathering is introduced as input parameter. It is assumed that dry deposition equals weathering.

2.2.2.2 Transformations of organic pools

A distinction can be made between primary and other organic pools. The primary pools are leaf litter and wood litter in the forest floor, and fine-root debris and coarse-root debris in the soil. These primary pools are fed by the corresponding vegetation components. The organic materials in the primary pools are subjected to conversion during one year. Of the woody pools only a fraction is susceptible to conversion. The conversion rates depend on temperature and chemical composition of the materials, which is characterized by their 'initial age' (Janssen, 1984). After a year the residues are transferred partly to the organic labile and partly to the organic moderately labile pool. The residence time in these pools is again one year, during which conversion continues but the rates are lower than for the primary pools. The residues of the organic labile pool are transferred to the organic moderately labile pool, and the residues of the organic moderately labile pool are transferred to the organic stable pool. The organic stable pool is converted at a still lower rate than the organic moderately labile pool.

The conversions of organic matter (carbon) consist of dissimilation and assimilation. By dissimilation organic matter is converted into H_2O and CO_2. By assimilation organic matter is converted into microbial tissue and this remains in the pool during the time step under consideration.

The conversions of organic nitrogen and phosphorus are related to the conversion of carbon. The assimilation of organic nitrogen and phosphorus by micro-organisms is proportional to the carbon assimilation. Those parts of converted organic nitrogen and phosphorus which are not assimilated in microbial tissue are mineralized, that is, added to the solution. If the C-N or C-P ratios of the organic pool exceed about 25 or 300, respectively, the mineralization is negative. In those cases some nitrogen or phosphorus is taken up by the microbes from the solution (immobilization).

Potassium is present only in the primary organic pools. A part of it leaches immediately from these pools to the solution, another part is gradually released during conversion, and the residues go directly to the solution. Thus there is no flow of potassium from the primary to the older organic pools.

All organic pools are prone to erosion, but their erosion rates may differ considerably. Of the woody primary organic pools yearly only the fraction that is susceptible to conversion is susceptible to erosion.

2.2.2.3 Nutrient flows to and from the solution

The soil solution is considered as a continuously stirred reactor. A number of simultaneous or sequential processes take place within each time step.

Processes by which the solution receives nutrients are:
- mineralization of nitrogen and phosphorus from the organic pools;
- atmospheric deposition of nitrogen and potassium;
- release from inorganic labile phosphorus;
- leaching and release of potassium from the primary organic pools;
- desorption of absorbed potassium.

Processes by which the nutrients leave the solution are:
- immobilization of nitrogen and phosphorus by the organic pools with a high C-N or C-P ratio;
- uptake by the vegetation;
- adsorption of potassium;
- leaching in case water percolates to the next deeper soil layer.

It is assumed that the plants are unable to take up nitrogen and phosphorus when the concentrations of these nutrients are below certain threshold values. The concentration of potassium follows from equilibrium with the exchange complex at the end of a time step.

2.2.2.4 Uptake of nutrients by and distribution in the vegetation

Only the nutrient with the highest uptake-supply ratio (Part II, Section 4) is completely used by the plant, taking into account the threshold concentrations for nitrogen and phosphorus. The uptake of the other nutrients is calculated as a function of the ratios of nutrient concentrations in the solution.

After the nutrients have been taken up, they must be allocated to the various plant components. A fraction of the absorbed nutrients is withheld in the fine roots. The remainder goes to the leaves. The absorbed quantity of a nutrient is calculated as a function of the available amount of that nutrient, regarding the possible maximum and minimum contents of the nutrient as boundary conditions. Nutrients not used by the leaves, if any, are available for wood growth. The wood receives nutrients mainly by translocation of nutrients from leaves shortly before they fall. Hence, shed leaves have a lower nutrient content than living leaves. Coarse roots receive nutrients via transformation of fine into coarse roots.

2.2.2.5 Growth

Growth is expressed as the increase in dry-matter mass of each plant component, per time step. Implicitly it is assumed that leaf growth is a result of photosynthesis, and that the growth of the other plant components is made possible by translocation and conversion of assimilates from the leaves to the other plant parts. Growth is the process by which carbon enters the ecosystem. The procedure for calculation of leaf growth is too complicated to summarize in a few sentences. The reader is referred to Part III, Section 3.4.3.2. The growth rate of leaves has a maximum that equals the quotient of the actual transpiration during that time step and the transpiration ratio.

For the other plant components the growth is calculated per time step as the ratio of the incoming quantity of the most limiting nutrient and the expected concentration of that nutrient. The expected nutrient concentrations for roots, stems and branches are related to the nutrient content in the leaves. (Part II, Section 5).

During each time step, fractions of the biomass of leaves, stems and branches, and coarse roots die and go to leaf litter, wood litter and coarse-roots debris (Fig. I.1). Fine roots loose a part of their mass by sloughing and another part by transformation into coarse roots.

3. REQUIRED DATA

To run the model a considerable number of data are required. It is often impossible to obtain these data. In such situations one has to rely on standard values and (pedo-) transfer functions.

3.1 Required climatological data

Rainfall
Usually, monthly rainfall over a number of years is used. Depending on the time step, daily data may be needed to calculate rainfall intensity. If only monthly (averaged) data are available, rainfall distribution must be generated (Van Diepen et al., 1988).

Pan evaporation
Also here monthly figures are used. Pan evaporation is translated into potential evapotranspiration by multiplying it with a 'crop factor'. For tropical forests the crop factor is calculated according to Poels (1987).

Temperature
For the calculation of conversion rates of organic pools, the average annual temperature must be known. In principle, it would be better to calculate conversion rates as a function of temperature, separately for each time step. This would require a tremendous quantity of calculation time. For tropical areas, where temperature hardly fluctuates from season to season, it is justified to use one value for temperature throughout the year.

3.2 Required litter data

3.2.1 Rates

Conversion rates
The assessment of conversion rates is based on the following equation from Janssen (1984, 1986):

$$Y_t/Y_0 = \exp \{4.7 [(a + f_t \cdot t)^{-0.6} - a^{0.6}]\} \qquad [I.1]$$

where:

Y_t = the amount of organic material remaining at time t

Y_0 = the initial amount of organic material

a = 'initial age' (years) of the organic matter; this can be considered as an index positively related to the resistance of the organic material against microbial conversion

f_t = temperature correction factor

t = time in years

If a and f_t are known, the average dissimilation constant (C_{dis}) over any time interval, Δt, can be calculated by: $C_{dis} = (1/\Delta t) \cdot \ln Y_t / Y_{(t+\Delta t)}$. Standard values of the 'initial age' are set at 2.18 for leaves, and 4.0 for wood. Depending on the resistance to conversion, higher or lower values for a can be applied.
The value of f_t can be found by

$$f_t = 2^{(T-9)/9} \qquad [I.2]$$

where T = temperature in °C.
For T = 27°C, f_t = 4; this is the value that can be used in many tropical forests. Under extreme conditions like waterlogging, drought, and very acid soils, dissimilation rates will probably be lower than calculated above. Corrections might be made by introducing more correction factors, e.g., f_m, correction for moisture content, and f_a, correction for acidity. So far, values for f_m and f_a have not been calibrated, and therefore these correction factors are not yet used in DYNAMITE.

Litter removal rates

It is assumed that only a fraction of dead wood and coarse roots is susceptible to conversion and to erosion in one year. For the Taï region and for Sarawak the value of the mean residence time was estimated at 15 years; this figure was derived from a study by Vooren (1985). For Sarawak, it proved necessary to assume that leaf litter too, has a mean residence time of more than one year; it was set at 3 years. In most cases, however, the mean residence time may be set at 1.0 year.

Mineralization of nitrogen and phosphorus

For the calculation of nitrogen and phosphorus mineralization, a method developed by Janssen and Noij (1993) has been incorporated in the model. Required data are:

- dissimilation-assimilation ratio for the conversion of organic materials; standard value is set at 2.
- C-N ratio of the micro-organisms which are involved in the conversion; standard value is set at 8.5.
- C-P ratio of these micro-organisms; standard value is set at 100.

Release of potassium
A fraction of potassium present in freshly fallen leaves is immediately leached; its value is set at 0.5.

3.2.2 State variables

Leaf litter
The initial quantities of organic matter, nitrogen, phosphorus and potassium are required as input data. If these data are not available, they can be estimated as a function of leaf mass, leaf fall rate and redistribution fractions, and conversion rates (Noij, 1988, p. 61).

Wood litter
Wood litter comprises both fallen and standing dead trees. Their quantity may be estimated by multiplying the number of such trees with their average mass. Also the quantity of fallen branches and twigs must be included in wood litter.

3.3 Required soil data

3.3.1 General

The number as well as the thickness of the soil layers must be known. The values may be different for the hydrological and the chemical (nutrient) part of the model.

Bulk density
For each soil layer the bulk density should be known to be able to translate data given per kg of soil into data per unit of volume (layer thickness times area).

3.3.2 Hydrological data

Maximum rate of capillary rise
If no data are available, they may be found by intra- or extrapolation of the data presented in Table III.1.

Water retention characteristics
Required are the volume fractions of moisture when the soil is:
- air dry;
- at permanent wilting point;
- at field capacity;
- at saturation.

Further it must be known at which volume fraction of air oxygen stress starts (Part III, Section 3.1.7)
If such data were not measured, they should be estimated from soil-texture analyses.

3.3.3 Chemical data

3.3.3.1 Rates

Wet deposition
The rates of wet deposition (nitrogen and potassium) can be calculated as the product of rainfall per time step and the concentration in rainfall. If such concentrations are not known they should be estimated from literature data (e.g., Poels, 1987).

Dry deposition
Data on dry deposition rate and on chemical composition of deposited materials are still harder to obtain than those on wet deposition. They must be estimated on the basis of existing literature and by common-sense interpolation (Stoorvogel, 1992). For the division of the dry deposition of inorganic phosphorus between inert and stable phosphorus, see below under 'Transformation of inorganic phosphorus' and Section 3.3.3.2 under 'Inorganic phosphorus'.

Erosion
Eroded materials often are relatively rich in organic matter, thus it can be concluded that relative erosion rates of organic pools are often higher than those of inorganic pools. Erosion under forest has seldom be measured. Values for relative erosion rate might be estimated from a range from zero to 0.002/year; of course they apply only to the upper soil layer.

Transformation of inorganic phosphorus
Inorganic inert phosphorus does not change by nature; it is assumed that the dry deposition equals the erosion of inorganic inert phosphorus. Values for the transformation fractions of inorganic labile and inorganic stable phosphorus are, for the time being, derived from Wolf et al. (1987) and Janssen et al. (1987). They are:

inorganic labile to solution	0.1/year
inorganic labile to inorganic stable	0.2/year
inorganic stable to inorganic labile	0.033/year.

Transformation rates of potassium
It is assumed that the following relationship exists between potassium in the solution ($SSOL_K$) and exchangeable potassium (ADS_K], both in kg ha^{-1}:

$$ADS_K = K_d \cdot SSOL_K \qquad [I.3]$$

where K_d is the adsorption constant; at present its value is set at 100.
This is a preliminary equation that can be applied as long as potassium is the only exchangeable cation considered in the model.
For the transformation of non-exchangeable (stable) potassium, so far no equations have been developed. For the time being, it is assumed that per hectare 20 kg of potassium are brought into the solution annually by weathering, and that an equal amount goes to the stable pool by dry atmospheric deposition.

Conversion rates of organic pools

The same formula applies as used for the calculation of the conversion rates of wood and leaf litter. The initial ages of the soil organic pools are set at:

- fine-root debris 2.18 year
- coarse-root debris 4.0 year
- labile organic pool 6.18 year (= age leaf litter + f_t)
- moderately organic pool 10.18 year (= age wood litter + f_t)

The relative dissimilation rate of the organic stable pool is found for a steady-state situation by solving the equation:

$$J_{tra,ORML} = ORST\ (1 - \exp(-C_{dis,ORST}/n)) \qquad [1.4]$$

where:

$C_{dis,ORST}$ = the dissimilation constant of organic stable pool (ORST)
$J_{tra,ORML}$ = rate of transfer from organic moderately labile to organic stable pool
n = number of time steps per year

Mineralization of nitrogen and phosphorus from the soil organic pools is calculated in the same way as for litter (Section 3.2.1), and hence no other data are required.

3.3.3.2 State variables

Organic carbon content

Soil organic carbon is present in three pools. It is not feasible, however, to measure individual pools. The magnitudes of the organic labile and the organic moderately labile pools follow from the sizes and conversion rates of the primary organic pools: leaf and wood litter, fine-root and coarse-root debris. Hence, the sizes of the organic labile and organic moderately labile pools can be found by running the model. To start such runs, as a first approximation it can be assumed that carbon is present in the organic labile, moderately labile and stable pools in proportions of 2 : 3 : 95.

Organic nitrogen content

Soil organic nitrogen is likewise present in three pools which cannot be measured individually. The distribution of organic nitrogen may be set at 1.5 : 2 : 96.5. It differs from the distribution of organic carbon because the pools differ in C-N ratios. These are approximately 17.3, 19.5 and 12.8 for the organic labile, moderately labile and stable pool, respectively, if the C-N ratio of total soil organic matter is set at 13. These values for C-N ratio can be used if soil organic nitrogen has not been analyzed. After running the model, final estimates for the distribution of soil organic nitrogen among the three pools can be made.

Organic phosphorus content

For soil organic phosphorus the same difficulties are met as for soil organic carbon and nitrogen. An additional problem is that values of soil organic phosphorus are rare, because the analytical procedure is more laborious and complicated than those for organic carbon and nitrogen. If no data are available, it is suggested to estimate soil organic phosphorus as a fraction of total P (range 0.3-0.7) and as the quotient of soil organic carbon and the C-P ratio (range for C-P ratio in tropical forest soils: 100-500). The ranges for soil organic phosphorus found with the two methods usually overlap; the middle of the overlap might be considered as the best estimate for soil organic phosphorus.

Inorganic phosphorus

Soil inorganic phosphorus is seldom determined and usually has to be estimated as the difference between total soil phosphorus and (the estimate for) soil organic phosphorus. The individual pools of inorganic labile (INLAP), stable (INSTP) and inert (ININP) phosphorus cannot be measured. They can be calculated for a steady state situation provided the rate of deposition of inorganic phosphorus, and the erosion and the transformation fractions of the inorganic pools are known. Deposition of inorganic inert phosphorus is then equal to the erosion of this pool. For the time being the values for the transformation fractions are derived from Wolf et al. (1987) and Janssen et al. (1987). Further the sum of the inorganic phosphorus pools must equal total inorganic phosphorus.

Using the values given in Section 3.3.3.1, the following equations can be derived for a time step of one year:

$$0.1 \cdot INLAP + 0.2 \cdot INLAP + FER_{INLAP} \cdot INLAP = 0.033 \cdot INSTP \qquad [1.5]$$

$$0.033 \cdot INSTP - FER_{INSTP}\ INSTP = 0.2 \cdot INLAP + ARDEP_{P} -$$

$$FER_{ININP} \cdot ININP \qquad [1.6]$$

$$INLAP + INSTP + ININP = TINP \qquad [1.7]$$

where:

FER_{INLAP} = erosion fraction of INLAP per time step
FER_{INSTP} = erosion fraction of INSTP per time step
FER_{ININP} = erosion fraction of ININP per time step

$ARDEP_P$ = annual rate of deposition of P (kg/ha per yr)
TINP = total inorganic P (kg/ha)

With these three equations, the values for the three unknown inorganic pools can be found, if FER_{INLAP}, FER_{INSTP}, FER_{ININP}, $ARDEP_P$ and TINP are known.

3.4 Required vegetation data

Relative rate of wood fall
This rate is given as a liner interpolation (AFGEN) function (Noij, 1988), p. 57, 59, 60). See Part III, Section 4. Original data were derived from Jaffré (1985), Jaffré & de Namur (1983) and Vooren (1985).

Relative rates of coarse-root and fine-root turnover
For the time being, for coarse roots this rate is arbitrarely set at two times the relative rate of wood fall, and for fine roots at 1.0/yr. No suited data were available (Part II, Section II).

Transpiration ratio
This is the ratio of transpired moisture and dry matter produced at moisture stress. If no appropriate data are available, its value is set at 300 kg/kg.

No other input data on rates in the vegetation are required, because all rates are calculated as functions availability of nutrients in the solution and the nutritional status of the vegetation. For this purpose, equations have been derived from literature discussed in Part II of this report.

Vegetation growth follows from model calculations. Hence, in principle no input data are required on biomass and nutrient contents of various vegetation components. The initial situation, however, should be described. If vegetation growth is simulated starting from bare soil, a certain minimum stand of the vegetation must be given. For this purpose, the values presented in Table I.1 can be used. If one wants to run the model starting from a standing vegetation, values for the variables mentioned in Table I.1 should be determined or estimated.

Table I.1 Initial values for biomass (dry matter) and nutrient contents of various components of a starting vegetation. All data are in kg per ha

Component	Biomass	N	P	K
Fine roots	100	1.0	0.15	1.0
Coarse roots	10	0.08	0.01	0.06
Wood	100	0.8	0.1	0.6
Leaves	100	2.5	0.3	2.0

4 EVALUATION

As mentioned in the introduction of this report, models serve to
- integrate the existing knowledge;
- improve the understanding of the functioning of the modeled ecosystem;
- indicate blank spots in our knowledge;
- assist in the planning of field research;
- predict the effects of (management) scenarios.

How is the performance of the present model? Before answering this question, it must be observed that DYNAMITE is still in development. These reports should therefore be considered as an account on the present state of the art.

4.1 Integration of existing knowledge

The existing knowledge might be distinguished into knowledge that has already been integrated in models, and knowledge that is scattered in reports, books and articles and not yet brought together in an assembly of operative calculation procedures.

Knowledge of the first type was exploited by incorporating (parts of) the following models:

ILWAS	Integrated Lake Watershed Acidification Study (Goldstein et al., 1984), including modifications made by Van Grinsven (1988);
WATERSTOF	Simulation of WATER and STOF (Wesselink, 1988);
SWATRE	Soil Water Actual Transpiration Rate Extended (Belmans et al., 1983; Feddes et al., 1978);
WOFOST	World Food Studies: crop growth simulation model (Van Diepen et al., 1988; Van Keulen and Wolf, 1986);
RESP	RESidual effect of Phosphorus fertilizers (Wolf et al., 1987);
QUEFTS	QUantitative Evaluation of the Fertility of Tropical Soils (Janssen et al., 1990);
MINNIP	MINeralization of NItrogen and Phosphorus (Janssen and Noij, 1993, Janssen 1984, 1986);
NUTCYC	NUtrient CYCling (Noij, 1988).

So far, most satisfying in DYNAMITE proved the calculation of the hydrological cycle, derived from ILWAS, WATERSTOF and SWATRE, and that of organic-matter decomposition and nitrogen and phosphorus mineralization, derived from MINNIP and NUTCYC. DYNAMITE deals with inorganic chemical processes still in a provisional way. Weathering processes have not yet been elaborated because of lack of time, but it is expected that ILWAS might provide useful techniques. Inorganic phosphorus reactions were described according to RESP. As this model has a time step of a year, its procedure is considered too crude for DYNAMITE.

Major difficulties were met with the vegetation compartment, especially with the distribution of nutrients among and the growth and dying of the different plant components. Some principles of WOFOST and QUEFTS could be used. A great portion of this part of the model, however, had to be newly devised, after a thorough examination of literature data (see Part II).

4.2 Improvement of understanding

Both models, NUTCYC and DYNAMITE, are suited to show how the growth of forest vegetation and the chances for substainable landuse are affected by
- dry and wet deposition;
- leaching of nutrients;
- (selective) erosion;
- sizes of nutrient pools in the soil;
- rates of decomposition and mineralization of organic pools;
- immobilization of nutrients during decomposition of woody materials;
- rates of wood and leaf fall;
- retranslocation (redistribution) of nutrients from leaves to wood.

The sensitivity of wood growth to some of these effects was demonstrated by Noij et al. (1988).

The surplus value of DYNAMITE above NUTCYC is that DYNAMITE set forth also:
- the interactions between moisture and nutrients and thus the effects of dry and wet spells and seasonallity in general on nutrient cycling;
- the interrelations among nutrients and thus the effects of e.g. balanced and unbalanced plant nutrition;
- the distribution of nutrients among above- and below-ground plant components and thus the effects of soil fertility on shoot-root ratios and on the ratio of fine to coarse roots;
- the effects of moisture and nutrient stress on the life span of leaves and thus on the rate of nutrient cycling;

- the effects of nutrient stress on nutrient-conservation mechanisms in the vegetation, such as retranslocation.

4.3 Indication of missing knowledge

It is general experience that during the development of a model questions arise that cannot be answered. The more sophisticated a model is, the more questions remain unanswered and the higher the number of 'guestimates' the modeler has to reconcile himself to. It is difficult to find universally applicable data, because the living community of a tropical forest, inclusive soil and litter fauna and microflora, proves unimaginably flexible and pliant in facing and circumventing adverse situations.

It does not make sense to exhaustively list the problems encountered. Some major questions were:
- the fraction of litter that is decomposed above ground and the fraction that is brought into the soil by animals or by leaching;
- the fraction of fallen and standing dead stems that is susceptible to decomposition and erosion;
- turnover rates of fine and coarse roots;
- dying rate of trees;
- the simultaneous occurrence or sequence of processes related to nutrients in the soil solution: microbial assimilation, uptake by the vegetation, adsorption and desorption, precipitation and dissolution, leaching;
- threshold nutrient concentrations in soil solution for uptake by vegetation;
- relation between nutrient ratios in soil solution and ratios of nutrients taken up by the vegetation;
- distribution of nutrients among plant components.

With respect to input data, it is often difficult to find reliable figures for:
- dry and wet deposition;
- rates of (selective) erosion;
- rates of mineral weathering;
- C-N and C-P ratios of micro-organisms;
- organic phosphorus content in the soil;
- contents of potassium in minerals.

Not yet considered in DYNAMITE are processes like:
- microbiological nitrogen fixation;
- leaking of nutrients from tree leaves;
- effects of pH and aluminum;
- denitrification and ammonia volatilization.

Relative erosion rates are at present input variables. It should be tried to relate them to weather (peak discharge) and physical soil data, so that it can become at output variable of the model.

4.4 Priorities in field research

The list of insufficiently known factors given in the preceding section points out for which topics field research would be required. Unfortunately, many of the mentioned items are difficult to measure, and that is of course the main reason why they have so seldom been assessed until now.

Not all questions, however, are equally important. Sensitivity analysis may be carried out to study the effects of variations for a number of variables. Often the interactions between some factors are more important than the effects of the factors individually. For instance, a low water-holding capacity of the soil is less a problem when rainfall is evenly distributed than when it is erratic.

It appeared (Noij et al., 1988) that for a given setting of climate, soil and vegetation characteristics, wood growth is very sensitive to the rates of atmospheric deposition and (selective) erosion, and to retranslocation of nutrients in the vegetation. On short term the initial size of soil state variables like phosphorus content is important, but on long term erosion and atmospheric deposition have much more effect.

From these results it can be concluded that, as far as field studies are concerned, priority should be given to the determination of dry and wet deposition of nutrients, erosion especially litter erosion, retranslocation of nutrients or generally nutrient conservation mechanisms in the vegetation. The establishment of C-P and C-N ratios in micro-organisms and the relationships between the ratios of nutrients in the solution and those of nutrients taken up, under different conditions, deserve first attention in laboratory and phytotron or greenhouse studies. Such studies should be preceded by a further scrutiny of literature.

Technically very difficult will be studies on root productivity and turnover, and distribution of nutrients among plant components. Nevertheless it is worthwhile to continue literature studies on these subjects, or to start such experiments if literature does not provide satisfying outcomes.

Further research is also required to find out when soil moisture content becomes critical for the uptake of the different nutrients, either by shortage of oxygen (wet soils) or in dry soils by increased tortuosity of the pathways the nutrients have to go to reach the roots. In the present version of the model, it is assumed that these moisture contents are the same as those for the reduction in water uptake. This simplication, however, is likely not justified.

4.5 Use of the model for prognostic purposes

Before the model is used as an instrument to predict the effects brought about by different management scenarios or different environmental conditions, it should first be validated by testing it in the field. This testing, however, is a complicated enterprise of long duration, because it requires at least repeated estimating of forest biomass over an extented period of time. In the ideal case, subroutines of the model are tested separately, but in practice this will usually remain a pious wish. The water balance as calculated with the present model was compared with data obtained by Poels (1987) in Suriname. The performance of this part of the model was very encouraging (Part III, Section 5.2).

Where testing is not possible or has not been done, it might yet make sense to run the model to investigate in what direction changes in management or environmental conditions will affect the growth of the vegetation. Examples of subjects that could be examined are:
- frequency of timber harvesting;
- shifting-cultivation intensity;
- the effect of gravel in the soil profile, for varying lengths of drought periods, on vegetation growth;
- the effect of peak discharges on nutrient output from the system and on vegetation growth.

Results of such studies without prior testing of the model are qualitative rather than quantitative, and should be interpreted with caution.

5 CONCLUSION

The model DYNAMITE is a tool to arrive at a better understanding of moisture and nutrient cycling in tropical forest lands. It should not be considered, however, as a final product. Testing the model in the field may result in considerable improvements. The authors hope that such field tests can be conducted at various sites of the Tropenbos Programme.

It is expected that DYNAMITE and its possible successors can make a substantial contribution to the design of effective measures for sustainable use of forest lands in the humid tropics.

REFERENCES

Anderson, J.M. & Ingram, J.S.I. (eds.). (1989). Tropical soil biology and fertility: a handbook of methods. CAB International, Wallingford, United Kingdom.

Andriesse, J.P. (1987). Monitoring project of nutrient cycling in soils used for shifting cultivation under various climatic conditions in Asia. (Final report). Royal Tropical Institute, Amsterdam, The Netherlands.

Belmans, C., Wesseling, J.G. & Feddes, R.A. (1983). Simulation model of the water balance of a cropped soil: SWATRE. J. Hydrol. **63**: 271-286.

Feddes, R.A., Kowalik, P.J. & Zaradny, H. (1978). Simulation of field water use and crop yield. Simulation Monographs. PUDOC, Wageningen, The Netherlands.

Goldstein, R.A., Gherini, S.A., Chen, C.W., Mak, L. & Hudson R.J.M. (1984). Integrated acidification study (ILWAS): a mechanistic ecosystem analysis. Philos. Trans. R. Soc. London, Ser. B 305: 409-425.

Jaffré, T. (1985). Composition minérale et stocks de bioéléments dans la biomasse epigée de recrûs forestiers en Côte d'Ivoire. Acta Oecoligica, Oecol. Plant., **6**: 233-246.

Jaffré, T., & de Namur, Ch. (1983). Evolution de la biomasse végétale epigée au cours de la succession secondaire dans le Sud-Quest de la Côte d'Ivoire. Acta Oecologica, Oecol.Plant., **4**: 250-272.

Janssen, B.H. (1984). A simple method for calculating decomposition and accumulation of 'young' organic matter. Plant and Soil **76**: 297-304.

Janssen, B.H. (1986). Een één-parametermodel voor de berekening van de decompositie van organisch materiaal. Vakbl. Biol. **66**: 433-436.

Janssen, B.H., Guiking, F.C.T., Van der Eijk, D., Smaling, E.M.A., Wolf, J. & Van Reuler, H. (1990). A system for quantitative evaluation of the fertility of tropical soils (QUEFTS). Geoderma **46**: 299-318.

Janssen, B.H., Lathwell, D.J. & Wolf, J. (1987). Modeling long-term crop response to fertilizer phosphorus. II. Comparison with field results. Agron. J. **79**: 452-458.

Janssen, B.H. & Noij, I.G.A.M., (1993). A simple model and a linear equation for the calculation of nitrogen and phosphorus mineralization. Plant and Soil.

Jordan, C.F. (1985). Nutrient cycling in tropical forest ecosystems. Wiley & Sons, Chichester, United Kingdom.

Mohren, G.M.J. (1987). Simulation of forest growth applied to Douglas fir stands in the Netherlands. (PhD thesis). Agricultural University, Wageningen, The Netherlands.

Noij, I.G.A.M. (1988). Een eenvoudig simulatiemodel voor nutriëntenbeperkte fytomassa produktie in tropische bosgebieden: NUTCYC. Verslagen en mededelingen 1988-1. Department of Soil Science and Plant Nutrition. Agricultural University, Wageningen, The Netherlands.

Noij, I.G.A.M., Janssen, B.H. & Van Reuler, H. (1988). Modeling nutrient cycling in tropical forest areas. A case study of Taï-region, Côte d'Ivoire. in: Proceedings Second International Tropenbos Seminar: 122-128. Tropenbos, Ede, The Netherlands.

Poels, R.L.H. (1987). Soils, water and nutrients in a forest ecosystem in Suriname. (PhD thesis), Agricultural University Wageningen, The Netherlands.

Sombroek, W.G. (1986). The need for common methodologies. Consultation on a united approach for research in the humid tropical forests. Tropenbos, Ede, The Netherlands.

Staritsky, I. (1988). Een model voor de nutriëntencyclus in het regenwoud in Sarawak. (Internal report). Department of Soil Science and Plant Nutrition, Agricultural University, Wageningen, The Netherlands.

Stoorvogel, J.J. (in prep.). Gross inputs and outputs of nutrients in undisturbed forest, Taï Area, Côte d' Ivoire. The Tropenbos Foundation.

Touber, L., Smaling, E.M.A., Andriesse, W. & Hakkeling, R.T.A. (1989). Inventory and evaluation of tropical forest land: guidelines for a common methodology. Tropenbos Technical Series **4**. Tropenbos, Ede, The Netherlands.

Van Breemen, N. (1988). Tropenbos: nutrient and moisture cycling, report of the working group. in: Proceedings Second International Tropenbos Seminar: 109-121. Tropenbos, Ede, The Netherlands.

Van Diepen, C.A., Rappoldt, C., Wolf, J. & Van Keulen, H. (1988). CWFS Crop growth simulation model WOFOST, (Documentation. Version 4.1). Centre for World Food studies, Amsterdam, The Netherlands).

Van Grinsven, J.J.M. (1988). Impact of acid atmospheric deposition on soils: quantification of chemical and hydrologic processes. (PhD thesis). Agricultural University Wageningen, The Netherlands.

Van Keulen, H. & Wolf, J. (eds) (1986). Modelling of agricultural production: weather, soils and crops. (Simulation monographs). PUDOC., Wageningen, The Netherlands.

Van Reuler, H. & Janssen, B.H. (1988). Soil fertility aspects of upland rice in shifting-cultivation systems in south-west Côte d'Ivoire. in: Proceedings Second International Tropenbos Seminar: 32-36. Tropenbos, Ede, The Netherlands.

Van Reuler, H. & Janssen, B.H. (1989). Nutritional constraints in secondary vegetation and upland rice in south-west Ivory Coast. in: Proctor, J. (ed.) Mineral nutrients in tropical forest and savanna ecosystems: 371-382. Spec. publ. no. 9 of the British Ecological Society, Blackwell Sci. Publ., Oxford, United Kingdom.

Vooren, A.P. (1985). Patterns in tree and branch-fall in a West African rain forest. (Report No. D 85-05). Department of Silviculture, Agricultural University, Wageningen, The Netherlands.

Wesselink, L.G. (1988). De St. Arnold lysimeters. Interpretatie van elementfluxen, simulatie van bodemhydrologie- en chemie. (Internal report). Department of Soil Science and Geology, Agricultural University, Wageningen, The Netherlands.
Wolf, J., De Wit, C.T., Janssen, B.H. & Lathwell, D.J. (1987). Modeling long-term crop response to fertilizer phosphorus. I. The model. Agron. **79**: 445-451.

PART II

NUTRIENTS AND PLANT GROWTH IN TROPICAL FORESTS

Literature study

CONTENTS

1 INTRODUCTION

1.1 Outline of the study

The purpose of this literature study is to gather data on ecophysiological relationships in tropical forests, and to interpret them for the purpose of modeling. The emphasis is on nutrients, and on aspects which had not or only provisionally been dealt with in the preceeding model NUTCYC.

The most important presupposition for the present model DYNAMITE is that the growth of humid tropical forests is mainly determined by the availability of nutrients. In Section 1.2 it is examined whether this presupposition is justified.

The uptake and distribution of nutrients strongly influence the production and distribution of dry matter in plants, and vice-versa. Therefore, it was necessary to study root production and turnover (Section 2) and dry-matter production and distribution in the above-ground parts of forest vegetations (Section 3).

The uptake of nutrients is a result of supply and demand. The quantities of nutrients which are supplied by the soil are not necessarily in balance with the demand of the plant, and hence the ratios of the absorbed nutrients often deviate from the ratios of the supplied nutrients. In Section 4, equations are derived for the calculation of the uptake of nutrients as a function of nutrient supply.

After the nutrients have been taken up, they must be distributed among roots, leaves, branches and stems. The procedure for nutrient distribution within plants is described in Section 5. It is based on the findings on dry-matter distribution discussed in Section 3 and on the relationships between nutrient concentrations in different plant components as derived from literature.

Tropical rain forests are well adapted to scarcity of nutrients. Retranslocation of nutrients from leaves to wood before abscission of the leaves is the most prominent nutrient-conserving mechanism. The fraction of the nutrients in the leaves that is retranslocated can be related to the nutrient status of the leaves (Section 6).

Nutrient status and moisture conditions influence both specific leaf area (leaf area per mass unit of dry matter), and leaf life span, and thus leaf area index (ratio of leaf area to land area). The leaf area index is a regulating factor for transpiration and evaporation. Regression equations for the calculation of specific leaf area as influenced by foliar nutrient contents, and for leaf life-span as a function of nutrient status and moisture stress are presented in Section 7.

Each section concludes with a set of calculation rules to be used in the model.

1.2 Is the growth of tropical forests nutrient-limited?

1.2.1 General remarks

The most important presupposition in NUTCYC, the precursor of the present model (Noij, 1988) was that plant growth in humid tropical forest areas is limited by phosphorus. However, it appears from literature that neither this presupposition, nor the presupposition that nutrients are limiting growth in these areas, is self-evident for the vegetation in forests.

Growth of annual crops tend to be nutrient-limited in these areas according to agricultural literature (e.g., Sanchez, 1976; Van Reuler & Janssen, 1988). Moreover, 63% of the soils in the humid tropics belong to the highly weathered and nutrient-poor Oxisols and Ultisols (Jordan, 1985). In contrast with annual crops, however, perennials and forest ecosystems possess a variety of nutrient-conserving mechanisms such as high nutrient efficiency (i.e. high dry-matter production per unit of nutrient), inherently low growth rates, high retranslocation efficiency, long leaf life-span, efficient nutrient-uptake mechanisms as a result of mycorrhizae, high fine-root turnover rates, and a root mat on the mineral soil. Some of these mechanisms contribute to maximizing the residence time of nutrients in the vegetation.

There is no agreement in the literature on the question of nutrient limitation of humid tropical forests (Jordan & Herrera, 1981; Proctor, 1983; Vitousek 1982, 1984), partly because data are insufficient to recognize tendencies, partly because it is hard to find locations which are differing in nutrient status only. To study the influence of several individual nutrients separately is even harder.

1.2.2 Nutrient concentrations in litter

In case of perennial plants, nutrient efficiency should be defined as the ratio of the amount of dry matter that is lost or permanently stored during a certain time period to the amount of nutrient that is lost or permanently stored during the considered time period (Vitousek, 1982).

In case root biomass is in steady state, and throughfall or nutrient leaking from the plant may be neglected, nutrient efficiency equals the reciprocal of the weighted average nutrient concentration of the phytomass increment and the produced dead material. In practice, this means that for a mature forest, nutrient efficiency is the reciprocal of the weighted average nutrient concentration of heartwood increment and litter production. Unfortunately, especially for the tropics, little is known about nutrient concentration of wood increment, but the few data available showed the same tendencies as Vitousek (1982) discovered for litter. Therefore he proposed to use production and nutrient content of litter as a measure for nutrient efficiency of the forest as a whole. This seems reasonable because only a small fraction of total above-ground nutrient uptake is assigned directly to the wood component (Section 5).

Vitousek (1982) found N efficiencies in litter varying from 45-250, P efficiencies varying from 800 (in one case 500) to 5000 and Ca efficiencies varying from 40-400 kg (dry matter) per kg of nutrient taken up, for a large number of forest ecosystems around the globe. The best correlation found was between litter production and the amount of N fallen with litter, which suggests N limitation in these forests. However, most tropical evergreen forests were not N-limited.

In a later publication, Vitousek (1984) reanalyzed the results for tropical forests. In this study, he related litter production to climatic factors. Deviations from the regression line between climatic factors and litter production proved to be correlated best with P and poorest with N. Hence, the highest correlation coefficient was found for the relation between litter production on the one hand and a combination of climatic factors and P concentration in the litter on the other hand ($R^2 = 0.60$; R^2 was 0.44 when P concentration was not included). The correlation coefficient was higher for the wet lowland tropics alone (altitude <500 m, precipitation > 1700 mm), and was even 0.94 for 'tierra firme' (i.e. never flooded) forest in the Amazon basin. Litter production probably was P-limited if the rate of litter-phosphorus turnover was less than 3 kg/ha per year and the P concentration in litter was less than 0.4 g/kg.
Comparable data for the cases were not found in literature. It is concluded that in wet lowland tropics P usually is the limiting factor, and in the temperate regions N is usually limiting. Transitional situations do occur.

2 ROOT DYNAMICS

2.1 Root dynamics and nutrient cycling

Production and turnover of roots, especially fine roots (d < 2, 5, 6 or 10 mm ; d stands for diameter), are hard to measure. The most obvious reason for this are the sampling problems. Apart from that, root production is not simply root increment as large amounts of small roots are continuously being sloughed during growth (Coleman, 1976). A fine-root sample therefore contains living and dead material, between which there is not always a clear distinction as illustrated by the term 'composite fine roots' (Klinge & Herrera, 1978). Most information on roots in forest ecosystems is therefore restricted to the standing crop of roots, and their distribution over diameter classes and through the soil profile.

Based on the scarce information available, many authors stress the importance of (fine-) root dynamics in nutrient cycling (Charley & Richards, 1983; Herrera et al., 1984; Jordan, 1985; Vitousek & Sanford, 1986). Turnover of fine roots appears to contribute more than 50% to the total annual nutrient turnover on several sites. Hence root dynamics should be included in a nutrient-cycling model and the best has to be made of the available information. Especially the roots finer than 2 mm require attention, because these roots are still unsuberized, have the highest nutrient absorption capacity and a very short life span (Newbould, 1968; Jenìk, 1978).

The growth and turnover of fine roots, their transition into coarse roots, and coarse-root turnover are processes which have implications for the model.

2.2 Roots in tropical forests

The available data on roots found for tropical humid forests (Table II.1) suggest a higher root-shoot ratio, a higher proportion of fine roots in total amount of roots, and a higher fine-root production rate on infertile than on fertile sites. A higher root-shoot ratio under nutrient-poor conditions is a well known phenomenon, both in agriculture (mostly monoculture) and in natural vegetation (Chapin III, 1980). Both high root-shoot ratio and higher proportion of finer root classes under nutrient stress may be the result of a higher (fine-)root production, a lower (fine-)root sloughing or both. In most cases fine roots have been measured without distinguishing living and dead roots. The relative decomposition rate of the fine-root debris may then affect the results which can be shown as follows. Let us assume that dead root material will not be recognized as such in a fine-root sample when it has been decomposed for more than one third and that the time required for such a decomposition is n days. For an estimation of n, we may apply Janssen's (1984, 1986) equations on "young" organic matter decomposition (Part I, Section 3.2.1). With an 'initial age' of 1.0 year and for f_t a value of 4.0 for the wet tropics, this means a dead root (d < 2 mm) will not be taken into account anymore in a sample,

Table II.1 Root characteristics of humid tropical forests arranged by soil fertility. After Vitousek & Sanford (1986), unless indicated otherwise

	Root-shoot ratio	Root biomass (Mg ha^{-1})			Fine roots		
Site		Total	Fine	Fine/ total	diam[o]., mm	production[p] Mg ha^{-1}	method
Moderately fertile soils							
1 Ghana, Kade[a]	0.11	22.1	4.45	0.20	6.25		
2 Costa Rica	0.04[b]	14.4[b]	2.9[c]	0.20	2	2.7[c]	k
Montane							
3 New Guinea	0.13	40.0[d]	2.8	0.07	5		
Infertile oxisols/ultisols							
4 Brazil	0.08	32.2	14.57	0.45	6		
5 Brazil[e]	0.26	255!	49!	0.19			
6 Venezuela	0.16	56	32[f]	0.57	6	2.0[g]	l
7 Venezuela[h]					2	15.4	m
Spodosols/psamments							
8 Venezuela	0.71	132.2	92.5[i]	0.70	6	32.9[j]	n
9 Brazil		29.5	15.9	0.54	6		

a Greenland & Kowal, 1960.
b Jordan, 1985.
c Raich, 1980.
d Edwards & Grubb, 1982.
e Klinge et al, 1975.
f Stark & Pratt, 1977.
g Jordan & Escalante, 1980.
h Sanford, 1985: upper 10 cm of the soil.
i Klinge & Herrera, 1978.
j Herrera & Klinge, 1979.
k Fine-root regrowth was measured 1 year after felling a 5-year old successional vegetation next to mature forest.
l Combination of three ingrowth experiments, lasting 9, 10 and 30 months.
m Monthly observation of sequential cores.
n Ingrowth experiment of unknown duration.
o Upper limit of diameter class.
p Production at the end of one year of ingrowth or regrowth.
! Fresh weight samples.

from 15 days after having been sloughed. Assuming a mean residence time of roots (d < 2 mm) of 122 days (see Section 2.3), the ratio of living to dead fine roots in steady state would then be 122/15 = 8.1, in other words about 11% of the fine roots measured would be dead. So, the interference of decomposition rates in the results of fine-root measurements for differing fertility levels, mentioned above, is probably not big.

2.3 Root turnover and root productivity

For temperate zones, more information on roots in forest ecosystems is available than for tropical zones (Vogt et al., 1986). For temperate, cold and boreal forests, fine-root turnover and mass are in the range of 0.5 - 16.0 Mg/ha per year, and 0.5 - 13.0 Mg/ha, respectively. The correlation Vogt et al. (1986) found between turnover and mass in cold-temperate and boreal coniferous forests suggests a mean residence time of fine roots (several classes) of about one year in these forests. For broad-leaved forests they found correlations between fine-root (several classes) turnover on the one hand and latitude ($R^2 = 0.61$, $P<0.10$) and mean climatic ratio (annual precipitation/mean annual temperature, mm/^{0}K; $R^2 = 0.79$, $P<0.05$) on the other. Extrapolation of these relationships to two exemplary tropical forests on the equator, both with a mean annual temperature of 300 ^{o}K (27°C) and an annual precipitation of 1500 and 3000 mm, leads to 'estimates' for fine-root turnover of about 10 and 40 Mg/ha per year, respectively. Although such an extrapolation is precarious, it gives an idea of the order of magnitude we may expect for fine-root turnover or productivity in humid tropical forests.

If root mass and root productivity are indicated by M and F (formation), respectively, and r is the relative rate of turnover, it holds:

$$dM/dt = F - r \cdot M \qquad [II.1]$$

If M = 0, at t = 0, integration yields:

$$M = (F/r)\,(1\text{-}\exp\,(-rt)) \qquad [II.2]$$

Applying this equation to three wheat experiments conducted to study root dynamics (Sauerbeck & Johnen, 1977; Martin & Puckridge, 1982; Keith et al., 1986), mean residence times were found of 45, 42 and 48 days. A faster turnover of wheat roots than of tree roots may be expected, because wheat has to take up nutrients and water within a much shorter growing season, and also because in case of wheat a larger fraction of the assimilates is available for root growth than in case of trees (no wood component, so less maintenance and conversion losses).

Sanford (1985, 1986) found for fine-roots (d < 2 mm) a relative turnover rate of 25% per month in a Venezuelan Oxisol with rain forest cover. This was based on monthly observations according to the sequential coring method, which is considered most reliable. This indicates a mean residence time of living roots (d < 2 mm) of 122 days, one third of the value mentioned above for cold regions. In view of the higher and more constant temperatures in tropical regions, resulting in higher and more constant physiological activity of fine roots, lower values for mean residence time may be expected in the humid tropics than in the temperate regions. The fact that the estimates of fine-root turnover are lower for tropical forest than for wheat indicates that the influence of plant species on root turnover is more important than the impact of climate.

Some other estimates of root turnover can be derived from the root-ingrowth experiments at Site 2 (Table II.1) for roots (d < 2 mm) and at Site 8 for roots (d < 6 mm), if it is assumed that the given values for root biomass refer to steady-state. The following equation applies for Site 2:

$$2.7 = (F/r)\,(1\text{-}\exp(-\,r \cdot 1)),\ \text{where } F/r = 2.9 \qquad [II.3]$$

Hence (1 - exp(- r)) = 0.931 and r = 2.674. The corresponding mean residence time is 136 days, which is close to the 122 days found by Sanford (1985) for roots (d < 2 mm).

A problem for Site 8 is that the authors do not mention how long the period of ingrowth was. Assuming a period of one year, it is found for Site 8: (1- exp (-r)) = 32.9/92.5 = 0.356, and r = 0.4395. The corresponding mean residence time would be 2.275 years. This value is unlikely high compared to the value of one third year for roots (d < 2 mm) and of one year for forest-tree roots in temperate regions. Perhaps the ingrowth period on Site 8 was less than one year.

It was decided to assume in the model that the mean residence time of roots (d < 6 mm) is one year, equal to the value in temperate regions. In the model roots (d < 6 mm) are not further subdivided into roots (d < 2 mm) and roots (2 mm < d < 6 mm), because it was impossible to find the required data.

2.4 Transition from fine to coarse roots

Fine roots disappear as such when they are sloughed, or when they grow thicker than 6 mm, or any other diameter that is taken as the boundary between fine and coarse roots. No records are known to us on the relative importance of the two processes. Estimates can be derived from the data of Table II.1, if some assumptions are made. The first one is that the data on dry matter of total and fine roots refer to steady-state situations. The second assumption, based on the fact that nutrient concentrations of coarse roots are lower than those of fine roots, is that the fine roots that pass into coarse roots form the only source of nutrients for coarse roots. If these two assumptions are correct, the loss of nutrients from dying coarse roots must equal the transfer of nutrients from fine to coarse roots. Thus it holds:

$$FCRTR \cdot GROOT_{DM} \cdot GROOTC_i = FFRTR \cdot FROOT_{DM} \cdot FFRCRT \cdot FROOTC_i \qquad [11.4]$$

where:

FCRTR	= fraction of coarse roots that is turned over per time step
$CROOT_{DM}$	= amount of dry matter in coarse roots
$CROOTC_i$	= concentration (mass fraction) of nutrient i in coarse roots.
FFRTR	= fraction of fine roots that is turned over per time step
$FROOT_{DM}$	= amount of dry matter in fine roots
FFRCRT	= fraction of rate of fine-root turnover that is transferred to coarse roots
$FROOTC_i$	= concentration (mass fraction) of nutrient i in fine roots

The transfer fraction (FFRCRT) can be found after reorganizing the above equation:

$$FFRCRT = (FCRTR/FFRTR)\,(FCROOT_{DM}/FROOT_{DM})\,(FCROOTC_i/FROOTC_i) \qquad [II.5]$$

FCRTR and FFRTR can be calculated as follows:

$$FCRTR = 1 - \exp(-\,RRCRTR \cdot \partial\ t) \qquad [II.6]$$

$$FFRTR = 1 - \exp(-\,RRFRTR \cdot \partial\ t) \qquad [II.7]$$

where:

RRCRTR	= relative rate of coarse-root turnover (1/yr)
RRFRTR	= relative rate of fine-root turnover (1/yr)
∂ t	= time step (yr)

For RRFRTR of roots (d < 6 mm), it was decided to use a value of 1.0/yr (Section 2.3). No data on RRCRTR are available. In the model it is assumed that RRCRTR is twice the relative rate of wood fall. In steady-state, relative wood-fall rate is assumed to be 0.01, and thus RRCRTR can be taken as 0.02. Hence, for a time step of one year, it holds:

$$FCRTR/FFRTR = (1-e^{-0.02})/(1-e^{-1}) = 0.0198/0.63212 = 0.0313 \qquad [II.8]$$

The values for $CROOT_{DM}$ and $FROOT_{DM}$ can be obtained from Table II.1. In Table II.3, $CROOT_{DM}$-$FROOT_{DM}$ ratios are given for five sites. Values vary from 0.43 (Site 8) to 3.97 (Site 1). Site 3 is further left out of consideration, because it has an exceptionally high value, namely 13.3.

Nutrient concentrations in roots are calculated in the model. Table II.8 in Section 5.5 presents minimum and maximum values of nutrient concentrations in the various plant components. In general, coarse roots have lower concentrations than fine roots. For the minimum and maximum concentrations the ratio $CROOTC_i/FROOTC_i$ was calculated (Table II.2). The value of the ratio increases in the order of N, P and K, and increases with increasing nutrient concentrations. The extremes are 0.3 and 1.0.

Table II.2 Ratios of the concentrations (g/kg) of N, P and K in coarse (d>6 mm) and fine (d<6 mm) roots, $CROOTC_i/GROOTC_i$, in the situations of minimum and maximum concentrations of N, P and K. See also Table II.6

	N	P	K
Minimum	0.30	0.42	0.80
Maximum	0.86	1.00	1.00

In Table II.3 the value of FFRCRT is calculated using Equation II.5, and 0.0313 and as value for FCRTR/FFRTR. This is done for the indicated ratios $CROOT_{DM}/FROOT_{DM}$ and for the two extreme values of $CROOTC_i/FROOTC_i$, being 0.3 and 1.0. The values calculated for FFRCRT range from 0.0040 to 0.1242, so a difference by a factor of 31. A still larger range might be expected if FCRTR/FFRTR is not kept constant. The ratio $CROOT_{DM}/FROOT_{DM}$ varies by a factor 10 and this is the main source of variation in the calculated FFRCRT values. The variation in $CROOT_{DM}/FROOT_{DM}$ is positively related to soil fertility.

When applying the model, the value of $CROOT_{DM}/FROOT_{DM}$ is an output and not an input parameter. Hence, it would be impossible to calculate FFRCRT with Equation II.5. Therefore, it is necessary to introduce in the model a parameter that is, like $CROOT_{DM}/FROOT_{DM}$, related to soil fertility. This parameter has been called Soil Fertility Index (SFI). It is proposed to use for SFI the ratio of actual (= nutrient limited) and potential (= water-limited) new growth of leaves. Both are calculated in the model for each time step (Part III, Section 3.4.3). The value of the ratio and thus of SFI lies between 0 and 1. In Equation II.5, SFI should take the place of $CROOT_{DM}/FROOT_{DM}$, of which the values range from 0.43 to 3.97 in Table II.3. So, the value of SFI cannot be equal to the value of $CROOT_{DM}/FROOT_{DM}$, in other words:

$$CROOT_{DM}/FROOT_{DM} = q \cdot SFI \qquad [II.9]$$

where q is a constant, estimated as follows. It was assumed that the maximum value of $CROOT_{DM}/FROOT_{DM}$ is equal to 5, a little higher than the value on Site 1 in Table II.3. This maximum will be obtained on fertile soils, where SFI equals unity. Hence:

$$q = (CROOT_{DM}/FROOT_{DM})/SFI = 5/1 = 5 \qquad [II.10]$$

It is obvious that further study is needed to check the relationships found so far to describe root dynamics.

2.5 Implications for the model

A fine (d < 6 mm) and a coarse (d > 6 mm) root class are distinguished. Nutrients for root growth are primarily assigned to fine roots. Coarse roots obtain nutrients only by transfer from fine to coarse roots. Transition of fine roots to coarse roots and sloughing are the processes by which fine roots disappear. The division of the relative turnover rate of fine roots over these processes is in the proportion of FFRCRT and (1-FFRCRT). The translocation fraction, FFRCRT, is:

$$FFRCRT = (FCRTR/FFRTR) \cdot q \cdot SFI \cdot (\text{min. of } CRNC_i/FRNC_i) \qquad [II.11]$$

where:

SFI = soil fertility index = actual new leaf growth/potential new leaf growth

$CRNC_i$ = the concentration (mass fraction) of nutrient i in the newly grown coarse roots

$FRNC_i$ = concentration (mass fraction) of nutrient i in the newly grown fine roots

i = N, P or K

min. = minimum value

The concentrations $CRNC_i$ and $FRNC_i$ are found as functions of leaf nutrient concentrations (Section 5.5).

Relative turnover rate of fine roots is set at 1.0/year, and that of coarse roots at two times the relative rate of wood fall. It is likely that the relative turnover rate of fine roots is related to soil fertility, but there are not enough data available to establish a relationship between soil fertility and the relative turnover rate of fine roots.

Table II.3 Ratio of dry matter in coarse roots ($CROOT_{DM}$) and fine roots ($FROOT_{DM}$) as derived from Table II.1, and translocation fraction (FFRCRT) for $CROOTC_i/FROOTC_i$ values of 1.0 and 0.3. For explanation see Section 2.4

Site	$CROOT_{DM}/FROOT_{DM}$	FFRCRT at $CROOTC_i/FROOTC_i$ of	
		1.0	0.3
1 Ghana	3.97	0.1242	0.0373
4 Brazil	1.21	0.0379	0.0114
9 Brazil	0.86	0.0269	0.0081
6 Venezuela	0.75	0.0235	0.0070
8 Venezuela	0.43	0.0135	0.0040

3 DRY-MATTER PRODUCTION AND DISTRIBUTION

3.1 Dry-matter production

In the nutrient cycling model DYNAMITE we started from the hypothesis that the production of vegetation is nutrient limited if no moisture stress occurs. This means that dry-matter production is not treated explicitly as a function of weather and canopy characteristics. In principle dry-matter production is calculated for every component in the vegetation as the increase in nutrient amount divided by the nutrient concentration (see Section 5 and Part III, Section 3.4). Yet, in order to get an idea on the order of magnitude of tropical-forest productivity and of the distribution of nutrients over above- and below-ground plant parts, the data of Table II.4 were gathered.

These estimates of primary productivity may be underestimates because (fine-)root productivity is generally not included (Jordan, 1985), or may be overestimates due to a bias in the selection of forests towards those relatively large in biomass and high in productivity (Brown & Lugo, 1984), and they may have been influenced by the inclusion of secondary forests. The first problem may be overcome by involving the root-productivity data of Section 2. Furthermore, it is assumed that Brown and Lugo's (1982) range is the most realistic for above-ground productivity, although it appears from the data of Jordan (1985) and Vitousek & Sanford (1986) that above-ground productivity lower than 10 Mg ha^{-1} yr^{-1} may occur on the less fertile sites.

Root formation (F) was calculated after reorganizing Eq. II.2:

$$F = M \cdot r/(1 - \exp(-rt)),$$

where:
M = 2.7 and r = 2.674 for Site 2,
M = 15.4 and r = 3.0 for Site 7 (Section 2.3)
t = 1 at both locations.

The root productivity at Site 7 may be higher than the calculated value of 48.6 tons/ha per year as only the upper 10 cm of the soil had been sampled.

Table II.4 Estimates for net primary productivity of tropical forests in Mg/ha per year

Forest type and/or component	Range*	Average	Reference
rain forests	10 - 35	22	a
seasonal forests	10 - 25	16	a
various forests	10 - 20		b
various forests			c
total	9.0 - 32.0		
leaves	2.1 - 12.0		
wood	4.6 - 19.2		
various forests			d
total		16.9	
leaves	s.d. = 7.2	9.6	
wood	s.d. = 5.6	7.3	
leaf litter	5.0 - 11.3		
wood	3.9 - 6.4		
various forests			
roots (d<2 mm)		7.75	e
roots (d<2 mm)		48.6	f

a Whittaker & Likens, 1975.
b Brown & Lugo, 1982.
c UNESCO, 1978.
d Jordan, 1985.
e Table II.1 of this study: Site 2.
f Table II.1 of this study: Site 7.
* s.d. = standard deviation.

3.2 Dry-matter distribution

From Jordan's (1985) data it may be derived that wood productivity is about 75% of leaf productivity. Above-ground productivity is assumed to be 10 Mg/ha per year and 20 Mg/ha per year at sites of low and high fertility, respectively. Other components than roots, wood and leaves are neglected (twigs, fruits, small branches) to calculate the typical dry matter distribution patterns of Table II.5.

The high total productivity at low fertility sites must be entirely ascribed to high root productivity. The apparent contradiction of high total productivity at low fertility level is a result of the fast turnover of fine roots. The low residence time of nutrients in the root compartment implies that in fact the same nutrients are used again and again to form new roots. As young roots are thinner, have a higher specific surface and are more actively absorbing nutrients than older roots, this phenomenon can be seen as an adaptive strategy to nutrient stress (Herrera et al., 1984).

Table II.5 Productivity of vegetation components in Mg/ha per year (A) and as fraction of total productivity (B) at sites of high and low fertility. Root productivity data from Table II.4

Components	Low fertility		High fertility	
	A	B	A	B
leaves	5.7	0.10	11.4	0.41
wood	4.3	0.07	8.6	0.31
(fine) roots	48.6	0.83	7.8	0.28
total	58.6	1.00	27.8	1.00

4 NUTRIENT UPTAKE

4.1 Nutrient ratios

The uptake of nutrients is a result of the supply by the soil and the demand by the plant. In DYNAMITE, the supply of nutrients is calculated as the net sum of a number of processes by which nutrients enter or leave the solution. The nutrients are supplied in ratios that often deviate from the optimum ratios for plant growth. As a consequence, nutrients are not taken up proportionally to their amounts in the solution.

In Section 4.2, nutrient ratios as found in leaves of tropical forests are compared with those found in annual crops. This is done to justify the application of results of pot trials with annuals for the quantitative description of nutrient uptake as a function of the supply of nutrients in the soil solution (Section 4.3). We had to use these data because no quantitative relationships between supply and uptake of nutrients by forest vegetation were found in the literature.

4.2 Ratios of nutrient concentrations in plants

The results of three pot trials were reinterpreted for the purpose of modeling. These trials, one with Chinese cabbage and two with maize, were conducted by students at the Department of Soil Science and Plant Nutrition of the Wageningen Agricultural Universtity (Rijkelijkhuizen, 1987; De Groof, 1988; Scheltema, 1989). All three experiments included different fertilizer combinations of N, P and K, thus creating strongly varying ratios in nutrient supply and uptake. The plants were harvested after about two months. From the results (Appendices II.2, II.3 and II.4) maximum and minimum ratios of nutrient concentrations were derived. They are compared with those in leaves of tropical forests (Table II.6). Also included are the maximum and minimum ratios found for the nutrients present in the above-ground plant parts of mature maize; these values were found in field experiments with maize in Kenya and Suriname (Janssen et al., 1990).

Table II.6 shows that the maximum N-P and the minimum K-N ratios of maize and Chinese cabbage correspond well with the respective ratios in tropical forest leaves. In natural environments, extreme supply ratios may be found only when nitrogen is involved. This is because in natural environments leguminous plants secure nitrogen supply to the vegetation (either directly or indirectly), leading to a relative abundance of nitrogen in otherwise nutrient-poor environments (high N-P and low K-N ratio). This implies that the minimum N-P and P-K, and the maximum P-K and K-N ratios of tropical forest leaves cannot be found without purposive experimentation. However, the available data do not suggest extreme nutrient ratios in tropical forest leaves entirely different from those found in the pot experiments. Therefore, it can be assumed that nutrient ratios of the above-ground plant parts in the pot experiments are the same as in leaves of tropical forests. This is considered

Table II.6 Minimum and maximum values of nutrient concentration ratios, in above-ground plant parts of Chinese cabbage, various plant parts of maize, and tropical-forest leaves

	N/P	P/K	K/N
Maximum ratios			
Chinese cabbage[a]	27	0.55	2.4
Maize, two months, pot trial[b]			
above-ground	30	0.67	2.9
roots	25	0.40	1.2
above-ground + roots[g]	23	0.75	2.3
Maize, mature, field trials[c]	20	0.60	2.3
Tree leaves			
Medina, 1984[d]	29.1	0.18	1.4
Appendix II.1[e]	36.7	0.16	0.9
Overall estimate	30	0.6	2.6
Minimum ratios			
Chinese cabbage[a]	3.0	0.035	0.28
Maize, two months, pot trial[b]			
above-ground	3.0	0.028	0.26
roots	6.0	0.036	0.26
above-ground + roots[g]	2.1	0.040	0.18
Maize, mature, field trials[c]	2.9	0.050	0.25
Tree leaves			
Medina, 1984[d]	10.7	0.044	0.35
Appendix II.1[f]	9.9	0.064	0.26
Overall estimate	3	0.03	0.26

a Rijkelijkhuizen, 1987.
b De Groof, 1988; Scheltema, 1989.
c Janssen et al., 1990.
d Data set of Medina (1984) includes several references from which extremes were chosen.
e Average of highest three values in Appendix II.1.
f Average of lowest three values in Appendix II.1.
g These values are not between those for roots and for above-ground parts, because they do not refer to exactly the same objects.

as an indication that the uptake ratios for forest vegetation do not substantially differ from those for annual crops.

In roots, the content of P is usually higher and the content of K lower than in the above-ground plant parts, and the range in nutrient ratios is smaller.

4.3 Relationships between nutrient supply and nutrient uptake

Uptake data for roots + above-ground plant parts of maize as found in the experiments by De Groof (1988) and Scheltema (1989) were used to relate nutrient uptake to nutrient supply. In both experiments, the potential nutrient supply by soil and fertilizers was calculated according to the procedure developed for the QUEFTS model (Janssen et al., 1990). Potential supply is defined as the uptake by the plant if no other growth factors than the considered nutrient are limiting. So, potential N-supply by the soil is the maximum N uptake by the plant at zero-N application, and the increase in potential N supply brought about by fertilizer N is the maximum amount of fertilizer N recovered by the crop. In trials set up to find the potential supply of a particular nutrient, usually rather high applications of the other nutrients are required.

The experiment by De Groof was a 3N x 5P x 5K factorial. Thus there were 15, 25 and 15 different ratios of the supplies of N and P, of P and K, and, of K and N, respectively. The experiment by Scheltema was a 3^3 factorial, resulting in 9 different ratios for each of the combinations N-P, P-K and K-N.

The experimental results have been elaborated in ratio diagrams, as introduced by De Wit (1960), and applied in plant-nutrition studies by e.g. Braakhekke (1980). First for each pot the supply and uptake ratios and their logarithms were determined. These will henceforth be denoted by log SN/SP, log SP/SK, log SK/SN, log UN/UP, log UP/UK, and log UK/UN. For each value of log SN/SP, the average of the corresponding values of log UN/UP was calculated. The same procedure was followed for log SP/SK and log UP/UK. The average values of the logarithms of the uptake ratios were plotted against the logarithms of the supply ratios (Fig. II.1 and II.2). It appeared that for a given log SN/SP, the values of log UN/UP were close together and were not affected by the level of SK. Similarly, log UP/UK did not vary much at a given log SP/SK, and was not affected by the level of SN. The values of log UK/UN, however, varied considerably at a given value of log SK/SN, and were strongly affected by the level of SP. Therefore, there was no purpose in seeking a relationship between log UK/UN and log SK/SN.
The relationships found in Fig. II.1 and Fig. II.2 are

$$\log UN/UP = 0.3 + 0.7 \log SN/SP \qquad [II.12]$$

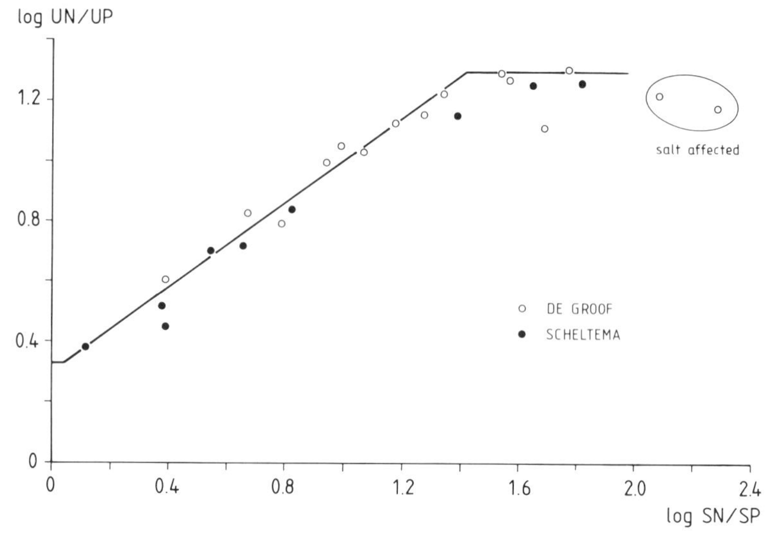

Fig. II.1 Relationship between the uptake ratio of nitrogen and phosphorus (UN/UP) and the supply ratio of nitrogen and phosphorus (SN/SP), for two-months old maize (De Groof, 1988; Appendix II.9; Scheltema, 1989; Appendix II.10).

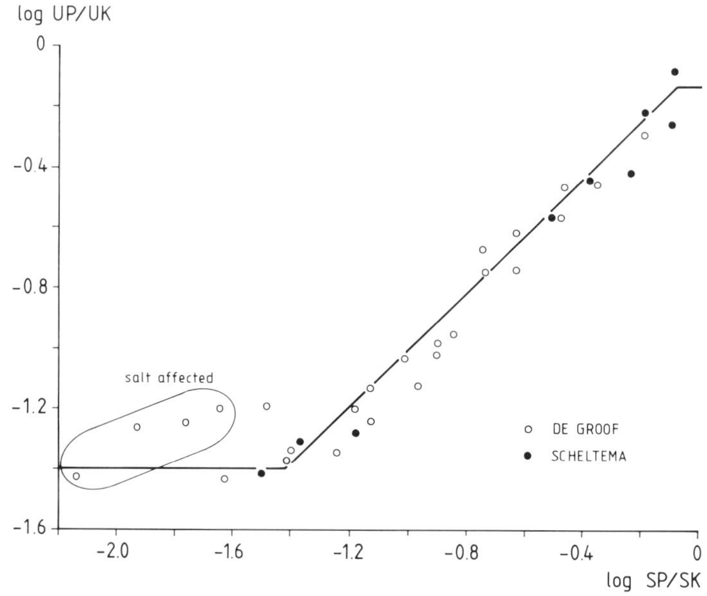

Fig. II.2 Relationship between the uptake ratio of phosphorus and potassium (UP/UK) and the supply ratio of phosphorus (SP/SK), and potassium for two-months old maize (De Groof, 1988; Appendix II.9; Scheltema, 1989; Appendix II.10).

$$\log UP/UK = -0.05 + 0.95 \log SP/SK \qquad [II.13]$$

For log UN/UP there seems to be a maximum value of 1.3, which means that the uptake of N will not be more than 20 times the uptake of P. For log UP/UK a minimum value of about -1.4 is found (Fig. II.2), which means that the uptake of P cannot be less than 0.04 times the uptake of K. In both cases, at the maximum value of log UN/UP and at the minimum value of log UP/UK, P was very strongly limiting plant growth and N and K fertilizers probably caused salt damage in these pots.

Equations II.12 and II.13 can be combined to formulate an equation for log UK/UN:

$$\log UK/UN = -0.25 + 0.95 \log SK - 0.7 \log SN - 0.25 \log SP \qquad [II.14]$$

Equation II.14 shows that with increasing supply of P, the value of UK/UN decreases. In other words, application of P stimulates the uptake of N more than the uptake of K, and this is exactly what has been observed in the pot experiments.

It cannot be seen from Fig. II.1 and Fig. II.2 whether there is a minimum value for log UN/UP or a maximum value for log UP/UK. The trials probably did not include a P level high enough to find such extremes. The maximum and minimum values mentioned in Table II.6 for these ratios can be used until more proper estimates have been found.

4.4 Implications for the model

The uptake of N, P and K is calculated according to the following rules. Equations II.12, II.13 and II.14 are applied to find UN/UP, UP/UK and UK/UN; maximum ratios are set at 20, 0.6 and 2.3, respectively, and minimum ratios at 2.1, 0.04 and 0.25, respectively.
It is assumed that the most limiting nutrient will be taken up completely. The uptake of the other nutrients follows from the uptake of the most limiting nutrient and the calculated uptake ratios. To identify the most limiting nutrient, the following ratios are calculated:

$$(UN/UP)/(SN/SP) = A \quad \text{or} \quad UN/SN = A \cdot UP/SP \qquad [II.15]$$

$$(UP/UK)/(SP/SK) = B \quad \text{or} \quad UP/SP = B \cdot UK/SK \qquad [II.16]$$

$$(UK/UN)/(SK/SN) = C \quad \text{or} \quad UK/SK = C \cdot UN/SN \qquad [II.17]$$

The coefficient with the highest value indicates the most limiting nutrient:
if A> B and C, N is most limiting;
if B> A and C, P is most limiting;
if C> A and B, K is most limiting.

5 NUTRIENT CONCENTRATIONS AND DISTRIBUTION

5.1 Method of data presentation

To obtain a better understanding of the way nutrient concentrations and distribution in natural vegetation vary among different sites in the humid tropics, data from several studies were compiled in Appendix II.1-II.7. The arrangement by soil fertility used there was adopted from Vitousek & Sanford (1986) who presented foliar nutrient concentrations in this way. For the same locations as for which these authors collected leaf data, it was tried to find as many data as possible on the nutrient concentrations of the other components of the vegetation. As data on nutrient contentrations are more scarce for wood and roots than for leaves, results from other studies were added to get a picture as complete as possible. A complication is that different authors distinguish or define vegetation components differently. In Appendices II.2 to II.7, this problem has been overcome by indicating with arrows to which component(s), each represented by a column, a certain number applies. If the boundary of a component had not been clearly defined, the point of the arrow has been omitted.

Vitousek & Sanford (1986) recorded foliar nutrient concentrations, generally calculated as the arithmetic mean of several species, whereas in our study geometric means were calculated for all components, if sufficient data were available. Geometric means are weighted for species abundance or for the biomass proportion of a particular vegetation component (e.g. undergrowth, palms, trees divided in DBH or D^2H classes). According to Tanner (1985) geometric means of foliar nutrient contents are usually lower than arithmetic means. This is true indeed for the Jamaican sites he studied, but not at many other places. Most authors present nutrient concentrations of a particular component for a clearly dominant vegetation component (e.g. trees with DBH > 10 cm). This implies that such data generally are fairly good estimates for the nutrient concentration of the component in the vegetation as a whole, and then the difference between arithmetic and geometric means are small. Therefore in case not sufficient data were available to calculate the geometric means, arithmetic means were included in our analysis.

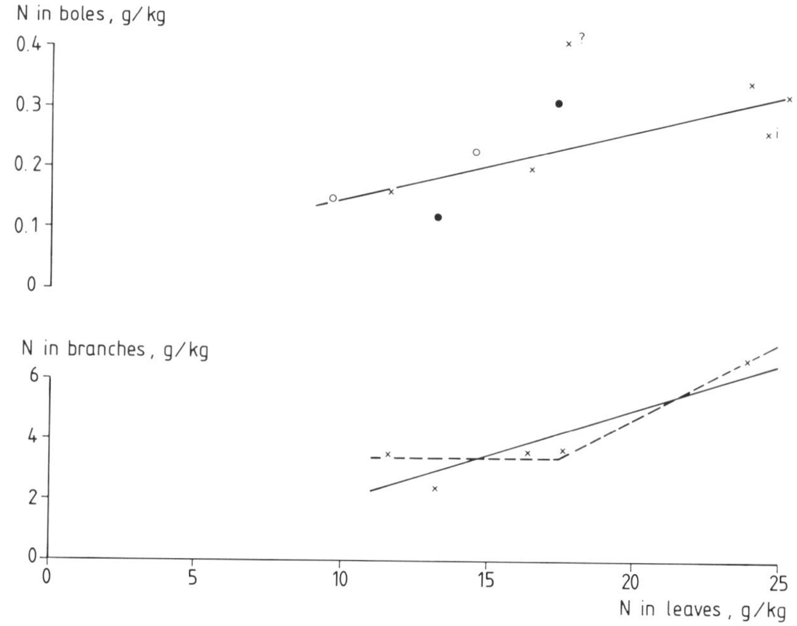

Fig. II.3 Relation between N-concentrations in leaves and boles (above), and leaves and branches (below) for several humid tropical forests (Appendices II.1 and II.2). Boles: o includes twigs and branches, . without bark, x no or only thickest branches included, ? left out for regression, i 18-year old forest. Dotted line: see text. Linear regression equations: see Table II.7.

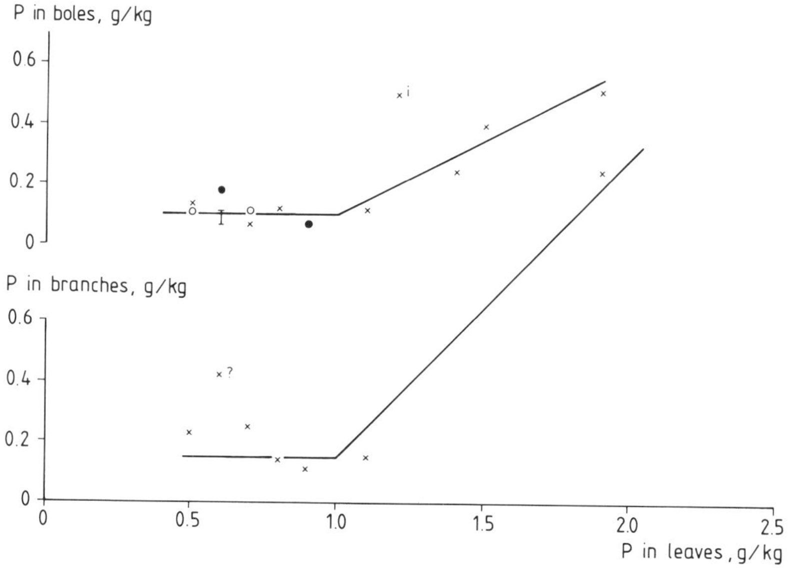

Fig.II.4 Relation between P-concentration in leaves and boles (above), and leaves and branches (below) for several humid tropical forests (Appendices II.1 and II.2). Boles: o includes twigs and branches, . without bark, x no or only thickest branches included, vertical lines represent range for heartwood - sapwood transition, ? excluded from analysis, i 18-year old forest. Linear regression equations: see Table II. 7.

5.2 Leaves

5.2.1 Nitrogen

Nitrogen concentrations in leaves range from 0.6 - 2.5 % for entire forests (Appendix II.1), whereas individual species may reach lower limits of 0.4 % (Medina, 1984). The highest levels are found on 'moderately fertile' soils, the lowest levels on 'Spodosols/Psamments' and some 'montane sites'.

Vitousek (1984) suggests that the vegetation on 'Spodosols/Psamments' is nitrogen limited. Indeed, lowest foliar N-P ratios and highest K-N ratios were found for 'spodosols/psamments' and one 'montane site' (Hawaii; Appendix II.1). Nitrogen levels are most probably correlated with conditions for nitrogen fixation in natural vegetation, unless there is some alternative import item like inundation water, such as for the igapo's in Brazil. Apparently there is no unambiguous relationship of such conditions with fertility level of the site, for foliar N concentrations are relatively high for 'infertile Oxisols/Ultisols' and vary widely on 'montane sites'. Probably, also factors such as climate, organic matter content, depth of soil profile and soil structure play a role. 'Montane sites' may have shallow soils and less favourable climates, 'Spodosols/Psamments' have an extremely low organic matter content, whereas 'Oxisols/Ultisols' generally have a deep soil profile, a good structure and a favourable climate.

5.2.2 Phosphorus

Foliar phosphorus concentrations range from 0.02 - 0.2 %, both for entire forests and for single species (Medina, 1984). The concentrations vary with soil fertility. The lowest values were recorded for sites that are known to have a low soil P availability: 'Oxisols/Ultisols', 'Spodosols/Psamments', and a poor montane site (Jamaica-mor). Not all 'Spodosols/Psamments' have low foliar P. Exceptions may be due to limitation by another nutrient, especially N (see Appendix II.1: N-P ratios). Although low P concentrations have been recorded for 'Oxisols/Ultisols', P-K ratios are not extremely low. This may also be the result of yet another limitation, probably cation availability (low K-N and ΣKCaMg-N ratios in Appendix II.1).

5.2.3 Potassium, calcium and magnesium

Foliar concentrations range from 0.4 - 1.9 % for K, 0.1 - 2.3 % for Ca, and 0.1 - 0.9 % for Mg (Appendix II.1). In general, the concentrations vary with soil fertility. The lowest values for K were recorded for Malaysia ('Spodosols/Psamments'), 'Oxisols/Ultisols', and Jamaica-mor ('montane') and the highest values for 'mode-

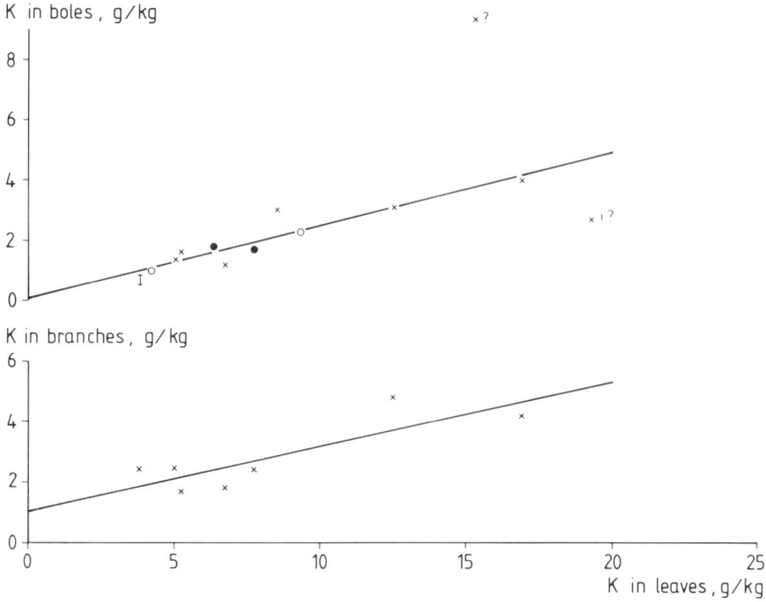

Fig. II.5 Relation between K-concentrations in leaves and boles (above), and leaves and branches (below) for several humid tropical forests (Appendix II.1 and II.2). Boles: o includes twigs and branches, . without bark, x no or only thickest branches included, vertical lines represent range for heartwood - sapwood transition, ? left out for regression, i 18-year old forest. Linear regression equations: see Table II.7.

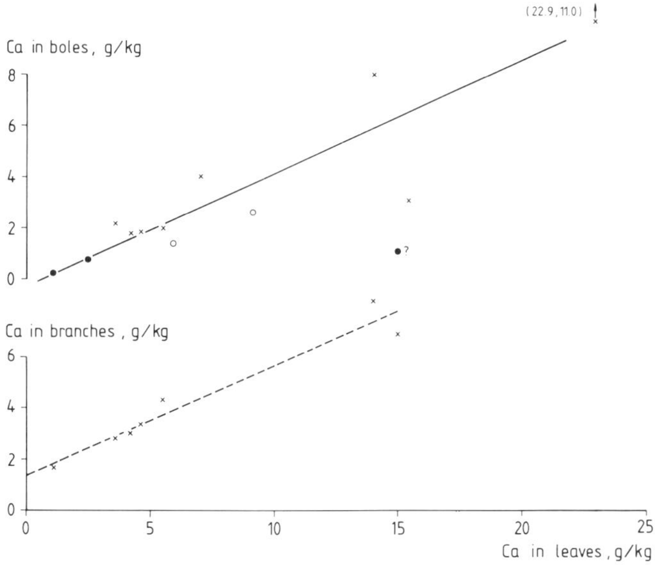

Fig. II.6 Relation between Ca-concentrations in leaves and boles (above),and leaves and branches (below) for several humid tropical forests (Appendices II.1 and II.2). Boles: o includes twigs and branches, . without bark, x no or only thickest branches included, vertical line represent range for heartwood - sapwood transition, ? left out for regression, linear regression. Linear regression equations: see Table II.7.

rately fertile soils'. The lowest values for Ca and Mg were found for the 'spodosol/psamment' the Igapó in Brazil and the 'Oxisols/Ultisols', and the highest values for 'moderately fertile soils'. As stated above, cations might be limiting on 'Oxisols/Ultisols', and K maybe also on some 'Spodosols/Psamments' and 'montane sites' (see nutrient ratios in Appendix II.1).

5.3 Stems, branches and twigs

5.3.1 Nitrogen

The branches, bark, sapwood and heartwood show N concentrations which are positively correlated to foliar concentrations (Fig. II.3). For branches and boles this relation may be described by linear regression. An alternative interpretation of the few data available for branches is that there is a constant N concentration of 3.5 g/kg in branches below a threshold value for the foliar N concentration of 17.5 g/kg, and that above this treshold the N concentration in the branches starts to increase (Fig. II.3, dotted line).

The three studies in which twigs were distinguished indicate a nitrogen concentration of about 0.8 % with little variation (Appendix II.2), although the foliar N concentrations ranged from 1.32 - 2.39 %.

Bark was analysed separately only twice. It had clearly higher N concentrations than the rest of the bole.

No study at all was found with separate data on N concentration for heart- and sapwood. Hase & Fölster (1982) found 0.34% N in the boles (Appendix II.2), but clearly lower N concentrations in the boles of the oldest and thickest trees (0.22 %). This indicates that N concentration in heartwood is lower than in sapwood, because the proportion of heartwood increases with the age of the tree. This is not confirmed, however, by Grubb and Edwards (1982, New Guinea), who found substantially higher N concentrations for older boles in nine out of ten species. Orman & Will (1960) and Bamber (1976), supplied evidence for withdrawal of N, P and K across the sapwood-heartwood transition in *Pinus radiata*, and of P and K in *Eucalyptus*. So, in two out of three studies N concentrations were lower in heartwood than in sapwood. The existing contradiction on this matter is no real problem for the growth module of the model, as no distinction between heart- and sapwood is made.

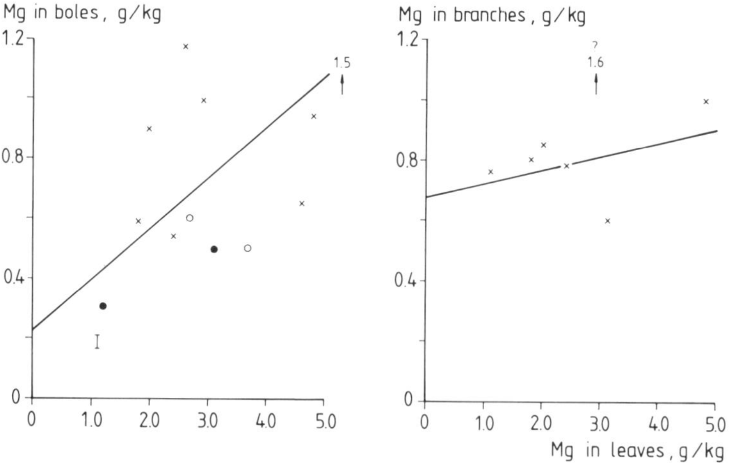

Fig. II.7 Relation between Mg-concentrations in leaves and boles (above), and leaves and branches (below) for several humid tropical forests (Appendices II.1 and II.2). Boles: o includes twigs and branches, . without bark, x no or only thickest branches include, vertical lines represent range for heartwood - sapwood transition, ? left out for regression. Linear regression equations: see Table II.7.

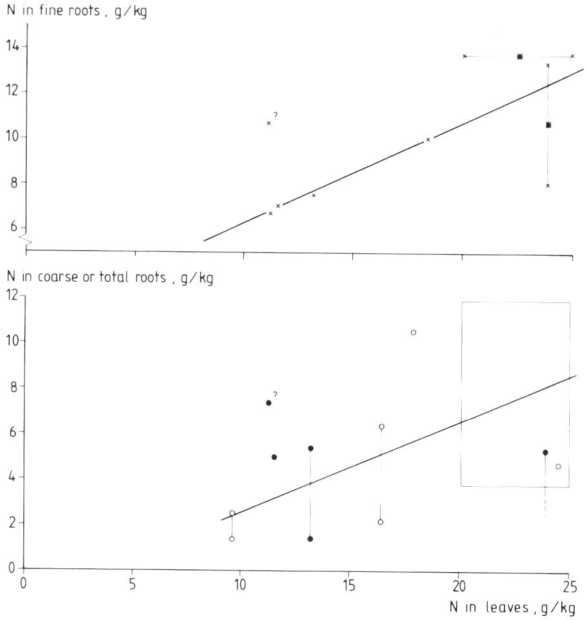

Fig. II.8 Relation between nitrogen concentrations in roots and leaves for several humid tropical forests (Appendices II.1 and II.3). Above: fine roots (d < 5, 6 or 6.25 mm). Below: . coarse roots (d > 5, 6 or 6.25 mm) and o total roots * average value used for regression, ? excluded from analysis. Vertical lines represent data of the same site for different root diameter classes, horizontal lines refer to an estimated leaf concentration range. Linear regression equations: see Table II.7.

5.3.2 Phosphorus

The concentrations of phosphorus in stems, branches and twigs are positively related to the corresponding foliar P concentration (Appendices II.1 and II.2; Fig. II.4). It may be concluded from these data that P concentrations in boles and branches start to increase, when the foliar concentration exceeds a threshold value of about 1 g/kg.

Except for Brazil - igapò, bark had higher P concentration than the rest of the bole. In the only study with separate data for heart- and sapwood (Venezuela-San Carlos) a higher P-concentration was found in sapwood. Hase & Fölster (1982, Venezuela) and Grubb & Edwards (1982, New Guinea), found lower P-concentrations in the boles of older trees compared to those in all trees (Venezuela: 0.025 % vs. 0.051 % in New Guinea; Appendix II.2). Together with the references cited in Section 5.3.1, these studies show a clear trend of lower P nutrient concentrations in heartwood than in sapwood.

5.3.3 Potassium, calcium and magnesium

The concentrations of cations in stems, branches and twigs appear to be positively related to those in the leaves (Appendices II.1 and II.2; Fig. II.5, II.6 and II.7). K and Ca concentrations in leaves show rather clear linear relationships with those in boles and branches. Data for Mg are much more scattered leading to low correlation coefficients.

The concentrations in bark are clearly higher than those in the rest of the bolewood, especially for Ca (5 to 20 times). From Grubb & Edwards' (1982) data it may be derived that bark constitutes about 10 % of the biomass of the bole. If this value also holds true elsewhere, and the concentration in the bark is ten times as high as the concentration in the rest of the bole, the concentration for the entire bole would be twice the value for the bole without bark (closed dots in Fig. II.6).

In Venezuela-San Carlos, the concentrations of K, Ca and Mg in sapwood are higher, somewhat higher and somewhat lower, respectively, than those in heartwood. Hase & Fölster (1982) found a lower K concentration, a lower Ca concentration and a higher Mg concentration, and Grubb & Edwards (1982) found higher K concentrations, lower Ca concentrations and lower Mg concentrations in old than in young boles. Although these data are ambiguous, a slight trend of Ca and Mg accumulation in heartwood and a light trend of K withdrawal from heartwood might be discerned, the latter being confirmed by Orman & Will (1960) and Bamber (1976); see also Section 5.3.1. Generally, it is found in plant nutrition studies that N, P and K are mobile, and that Ca is not mobile, while Mg takes an intermediate position (Mengel & Kirkby, 1979.)

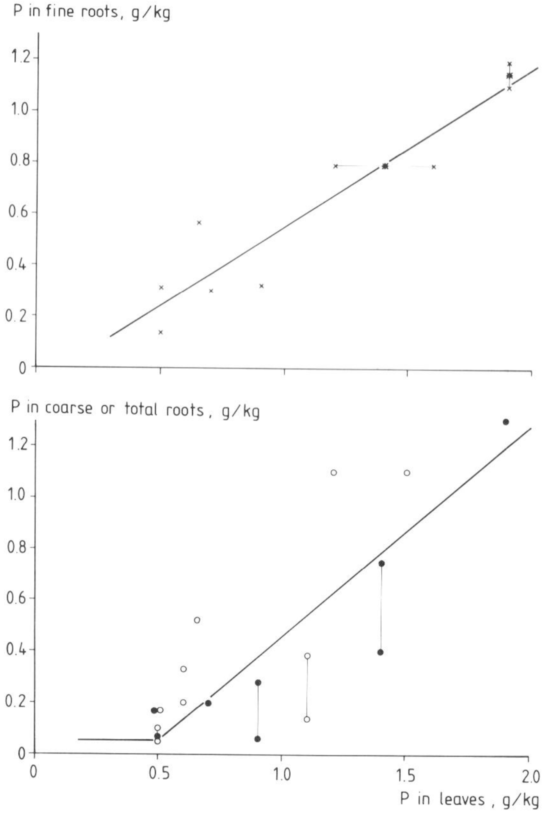

Fig. II.9 The relation between phosphorus concentrations in roots and leaves for several humid tropical forests (Appendices II.1 and II.3). Above: fine roots (d < 5, 6 or 6.25 mm). Below: . coarse roots (d > 5, 6 or 6.25 mm) and o total roots. * average value used for regression, ? excluded from analysis. Vertical lines represent data of the same site for different root diameter classes, horizontal lines refer to an estimated leaf concentration range. Linear regressions equations: see Table II.7.

5.4 Roots

5.4.1 Nitrogen and phosphorus

A straight line seems to be a reasonable description of the relation between N or P concentrations in roots (d<6 mm) and in leaves (Fig. II.8 and II.9). Concentrations in roots (d>6 mm) or total roots are generally lower compared with roots (d<6 mm), except for P at high foliar concentration. As a rule, concentrations decrease with increasing root diameter (Appendices II.3 and II.4), but there are some exceptions. In Ghana-Kade, New Guinea and Jamaica-mor, roots of the coarsest classes had a higher P concentration than those of one class finer. This may be interpreted as P storage in the coarsest roots.

5.4.2 Potassium, calcium and magnesium

Data on cations are erratic (Appendices II.5, II.6 and II.7; Fig. II.10, II.11 and II.12) compared to those on N and P. The correlation between fine-root and leaf concentrations is rather poor, but clearly positive. As a consequence, the interpretation of the data was rather arbitrary. Further studies will have to confirm or refute the lower and upper limit suggested for K and Ca concentrations in fine roots in Fig. II.10 and II.11. Coarse-root concentrations seem also to be positively correlated too with foliar concentrations. It is hardly possible to distinguish between fine and coarse roots in this respect, but K and Ca concentrations in coarse roots tend to be somewhat higher than those in fine roots at low foliar concentrations.

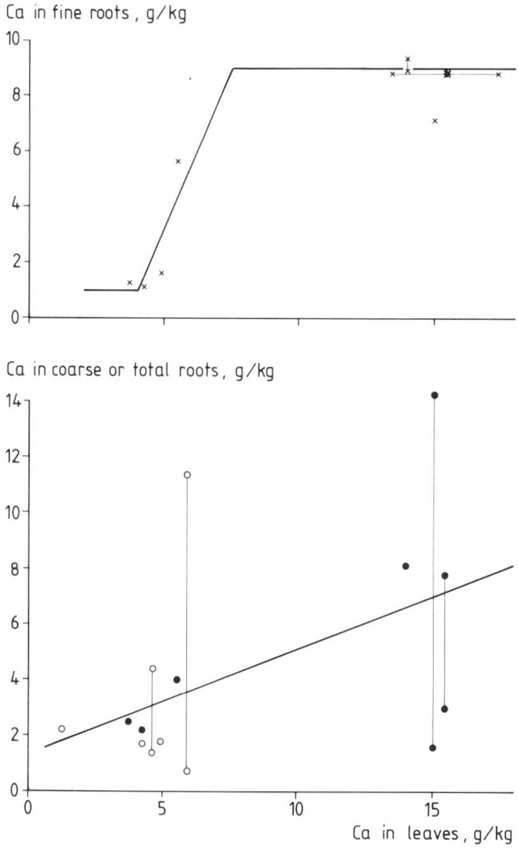

Fig. II.11 Relation between calcium concentrations in roots and leaves for several humid tropical forests (Appendices II.1 and II.3). Above fine roots (d < 5, 6 or 6.25 mm). Below: . coarse roots (d > 5, 6 or 6.25 mm) and o total roots. * average value used for regression, ? excluded from analysis. Vertical lines represent data of the same site for different root diameter classes, horizontal lines refer to an estimated leaf concentration range. Linear regression equations: see Table II.7.

Table II.7 Relationships between nutrient concentrations (g/kg) in boles, branches or roots (y), and in leaves (x)

Nutrient	Equation	R^2	Corresponding figure
Boles			
N	y = 0.34 + 0.115 x	0.673	II.3
P	if x < 1 , y = 0.1 else		II.4
	y = -0.4 + 0.5 x	n.c.[a]	
K	y = 0.07 + 0.243 x	0.876	II.5
Ca	y = - 0.25 + 0.440 x	0.798	II.6
Mg	y = 0.23 + 0.169 x	0.369	II.7
Branches			
N	if x <10, y = 2		II.3
	y = - 1.0 + 0.31 x	0.799	
P	if x < 1 , y = 0.15, else		II.4
	y = - 0.85 + x	n.c.	
K	if x > 18.5, y = 5.0, else		II.5
	y = 1.06 + 0.213 x	0.843	
Ca	y = 1.38 + 0.426 x	0.945	II.6
Mg	y = 0.68 + 0.046 x	0.224	II.7
Roots (d < 6 mm)			
N	y = 1.86 + 0.441 x	0.822	II.8
P	y = -0.7 + 0.622 x	0.860	II.9
K	if x < 5.1 , y = 1		II.10
	if x > 10.3 , y = 7.5, else		
	y = - 5.39 + 1.255 x	0.688	
Ca	if x < 4 , y = 1		II.11
	if x > 7.5 , y = 9 , else		
	y = - 8.01 + 2.269 x	0.686	
Mg	y = 0.51 + 0.351 x	n.c.	II.12
Roots (d > 6 mm)			
N	y = - 1.4 + 0.4 x	n.c.	II.8
P	if x < 0.5, y = 0.05, else		II.9
	y = - 0.36 + 0.82 x	n.c.	
K	if x < 4.6, y = 0.8		II.10
	if x > 18, y = 7.5, else		
	y = - 1.5 + 0.5 x	n.c.	
Ca	y = 1.375 + 0.375 x	n.c.	II.11
Mg	y = 0.42 + 0.25 x	n.c.	II.12

[a] n.c. = not calculated because the number of data was too restricted; see corresponding figures.

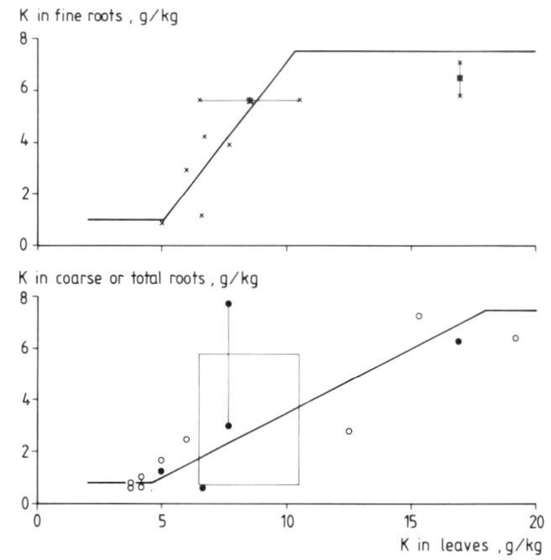

Fig. II.10 Relation between potassium concentrations in roots and leaves for several humid tropical forests (Appendices II.1 and II.3). Above: fine roots (d < 5, 6 or 6.25 mm). Below: . coarse roots (d > 5, 6 or 6.25 mm) and o total roots. * average value used for regression, ? excluded from analysis. Vertical lines represent data of the same site for different root diameter classes, horizontal lines refer to an estimated leaf concentration range. Linear regression equations: see Table II.7.

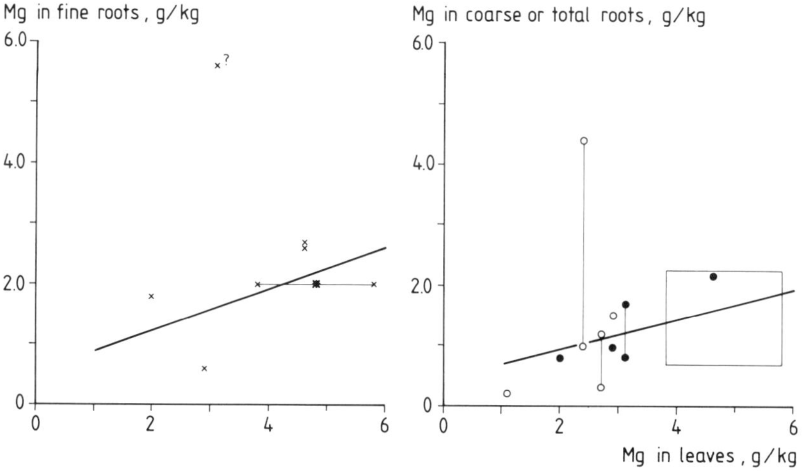

Fig. II.12 Relation between magnesium concentrations in roots and leaves for several humid tropical forests (Appendices II.1 and II.3). Above: fine roots (d < 5, 6 or 6.25 mm). Below: . coarse roots (d > 5, 6 or 6.25 mm) and o total roots. * average value used for regression, ? excluded from analysis. Vertical lines represent data of the same site for different root diameter classes, horizontal lines refer to an estimated leaf concentration range. Linear regression equations: see Table II.7.

5.5 Implications for the model

In the model, N, P and K concentrations in wood and roots are calculated as a function of the respective foliar concentrations according to the relationships found in this chapter. These relationships are described by the equations in Table II.7 which also contains data on Mg and Ca, for the sake of completeness.

Minimum and maximum concentrations in the various plant components are presented in Table II.8. These have been derived from Figures II.3 - II.5 and II.8 - II.10 and the corresponding appendices (Appendix II.1 -II.5). The data of Tables II.7 and II.8 are used to estimate the distribution of absorbed nutrients among the plant components. The exact calculation procedure is given in Part III, Section 3.4.3. It is based on the principles described below.

The absorbed nutrients are distributed in the first instance over fine roots and leaves; in formula:

$$UP_{act,i} = GROWFR \cdot FRNC_i + GROWL \cdot LNC_l \qquad [II.18]$$

where

$UP_{act,i}$	= total uptake of nutrient i
GROWFR	= dry matter in newly formed fine roots
GROWL	= dry matter in newly formed leaves
$FRNC_i$	= concentration of nutrient i in newly formed fine roots
LNC_i	= concentration of nutrient i in newly formed leaves

The known factors in this equation are the total uptake of nutrients (see Section 4.4) and the concentration of nutrients in fine roots, the latter being calculated as a function of nutrient concentration in leaves at the beginning of the time step. The unknown factors are GROWFR, GROWL and LNC_i. Now it is assumed that the concentration of the growth limiting nutrient will be at minimum in the leaves. So, this concentration, indicated by $LEAFCD_{il}$, is also known. To determine which nutrient is yield limiting, the ratios $UP_{act,i}/FRNC_i$ are calculated. The nutrient with the lowest ratio is considered to be yield limiting. Now GROWFR can be expressed as a function of GROWL:

$$GROWFR = UP_{act,i}/FRNC_i - GROWL \cdot LEAFCD_{il}/FRNC_{il} \qquad [II.19]$$

$$GROWFR = A - B \cdot GROWL \qquad [II.20]$$

Table II.8 Minimum and maximum concentrations of nutrients (g/kg) in various plant components

Component	Nutrient	Minimum	Maximum
Leaves	N	7.5	25
	P	0.3	2
	K	4.0	20
Boles	N	1.2	3.2
	P	0.1	0.6
	K	1.0	5.0
Branches	N	2.0	7.0
	P	0.15	1.15
	K	1.9	5.0
Roots, d<6 mm	N	5.0	14.0
	P	0.12	1.3
	K	1.0	7.5
Roots, d>6 mm	N	1.5	12.0
	P	0.05	1.3
	K	0.8	7.5

Equation II.20 is substituted in the equation for $UP_{act,i}$, (Eq. II.18), where i refers to the non-limiting nutrients.

$$UP_{act,i} = (A - B \cdot GROWL) \cdot FRNC_i + GROWL \cdot LNC_i \qquad [II.21]$$

Apart from GROWL, LNC_i is unknown in this cquation. The maximum and minimum values of LNC_i, however, are known (Table II.8). For each of the two non-limiting nutrients, subsequently the maximum ($LEAFCA_i$) and the minimum value ($LEAFCD_i$) is substituted, resulting in two estimates of GROWL: $GROWLA_i$ and $GROWLD_i$ (A stands for accumulated and D for diluted). The new growth of leaves is finally calculated as the middle of the common overlap of the two yield ranges between $GROWLA_i$ and $GROWLD_i$. If there is no overlap GROWL is assumed to equal the lower of the two $GROWLD_i$ values.

Once GROWL is known, GROWFR can be calculated, and subsequently the nutrients required in the new fine roots and leaves. If total uptake exceeds the quantities needed for fine roots and leaves, the overflow is sent to wood.

6 NUTRIENT RETRANSLOCATION

6.1 Reabsorption efficiency

Retranslocation of nutrients within the plant, sometimes called biochemical cycling, is an important process in nutrient cycling. It may satisfy about 50% of the demand of N and P and 20% of the demand of K, Mg and S in forests. However, it hardly contributes to the satisfaction of Ca demand (Charley & Richards, 1983). The most important redistribution of nutrients is that from leaves before abscission. Although translocation of nutrients from sapwood to heartwood deserves attention too, it is not treated here as no distinction is made between these two wood compartments in the model.

Estimates for the reabsorption efficiency (i.e. the retranslocated fraction of nutrients present in the living leaf) for N and P range from 35 - 80% for a variety of natural vegetations including tropical forests (Medina, 1984; Vitousek & Sanford, 1986; Lajtha, 1987). The estimate for K ranges from 10 - 20% for northern hardwood forests (Ryan & Borman, 1982), while no or hardly any redistribution of Ca is recorded. According to Chapin III (1980) the fraction of the maximum leaf content that is reabsorbed ranges from 50 to 90% for N and P, and is below 70% for K. Redistribution of K is difficult to measure because the element easily leaks from both living and dead leaves. We rely on the range of Ryan & Borman (1982) with an average of 15 % for K-reabsorption efficiency.

6.2 Nutrient status and reabsorption efficiency

The information in the literature on the influence of the nutrient status of natural vegetations on reabsorption efficiency appears contradictory. Chapin III (1980) concludes that.. "In general, plants of high nutrient status retranslocate a larger proportion of their leaf nitrogen and phosphorus than plants with low nutrient status." Nevertheless, differences in nutrient concentration between living leaves and leaf litter may be greater on infertile sites due to different weight loss prior to abscission. However, for tropical forests weight losses calculated from the change in Ca content range from 12 (montane) to 37.5% (infertile Colombian oxisol) with an average of 28%, without any consistent relation to the fertility level of the sites (data taken from Table 5 in Vitousek & Sanford, 1986; for the most fertile site Ghana in this data, 23% was calculated). So, the reabsorption efficiency in tropical forests can probably be derived from, and is positively related to the difference in nutrient concentration between living and abscised leaves.

Vitousek (1982, 1984) studied production and nutrient content of litter for a wide range of fertility levels. We applied the following assumptions to his data:

- N and P content in living leaves ranged from 10 - 25 and 0.4 - 2.0 g/kg, respectively (Table II.8);
- litter consists for 70% of leaf litter (Proctor, 1983);
- compared with leaf litter, the remainder 30% of the litter has half the N concentration in all cases, one third the P concentration under infertile conditions and equal P concentration under fertile conditions (compare 'twigs' with 'leaves' in Appendices II.1 and II.2).

In Vitousek's data N concentrations in litter ranged from 6 - 19 and P concentrations from 0.17 to 1.4 g/kg (excluding the most extreme low values), yielding estimates for N concentrations in leaf litter ranging from 7.1 to 22.4 and from 0.21 to 1.4 g/kg for P concentrations. If a weight loss during abscission of 28% (see above) is assumed, reabsorption efficiency would range from 49 - 36% N and 62 - 50% P from low to high fertility level, that is a trend opposite to the one suggested by Chapin III (1980).

Lajtha (1987) cites various authors who have found decreasing reabsorption efficiency with increasing nutrient availability, and various others who have found no such relation or the reverse. She argues that 'Previous workers may have reached contradictory interpretations regarding the response of plants to nutrient stress because the status of experimental sites along the full gradient of nutrient availability was unknown'. She suggests a model of the relationship between nutrient availability and nutrient reabsorption efficiency (Fig. II.13). At low nutrient availability the absolute amount of nutrients available for reabsorption is limited, because less soluble and/or hydrolyzable compounds containing the nutrient are present, whereas in more fertile sites the cost of reabsorbing a nutrient may exceed the cost of nutrient uptake by the roots.

The values for the reabsorption efficiency found by Vitousek & Sanford (1986) were plotted against the corresponding nutrient concentration in the living leaves (Table II.9, Fig. II.14: dots) together with the ranges calculated above (Fig. II.14: crosses).

Figure II.14 shows that the values calculated by these authors agree with the ranges estimated above and that the model suggested by Lajtha (1987) as represented by the solid lines a and b, and the dotted lines possibly applies. The dotted lines were arrived at as follows.

According to Chapin III (1980), 10% of the maximum content of N and P in the leaves of natural vegetation cannot be redistributed, i.e. about $0.1 \cdot 25 = 2.5$ g/kg N and $0.1 \cdot 2 = 0.2$ g/kg P in forest leaves. The data of Vitousek (1982, 1984) suggest minimum concentrations in leaf litter of 7.1 and 0.21 g/kg for N and P, respectively. Assuming a weight loss of 28% during abscission, this means the N and P concentration in litter, expressed as mass fraction of the original leaf dry matter, would be $7.1 \cdot 0.72 = 5.1$ g/kg for N and $0.21 \cdot 0.72 = 0.15$ g/kg for P. When these fractions are compared with the theoretical ones from above, derived from Chapin III (1980), it appears that tropical forests redistribute P more efficiently than N. If it is assumed that all N or P above the minimum concentrations expressed on the basis of original weight (5.1 and 0.15 g/kg for N and P, respectively) can be redistributed, the dotted lines of Fig. II.14 are the result.

Table II.9 Foliar N and P concentration (g/kg) and reabsorption efficiency (%), and N-P ratio of the redistributed material. After Vitousek & Sanford (1986). If indicated by*, after Fölster et al. (1976)

Site	Concentration		Reabsorption efficiency		N/P
	N	P	N	P	
Moderately fertile					
Ghana	25.2	1.4	36	51	2.7
Infertile Oxisol/Ultisol					
Colombia*	17.6	0.8	54	73	16.4
Spodosols/Psamments					
Venezuela, San Carlos	17.8	0.6	42	67	18.6
Venezuela, Caatinga	10.8	0.7	55	50	17.0
Venezuela, Open Bana	8.9	0.4	44	56	17.5
Montane					
Venezuela	17.4	0.8	40	32	27.2
Papua N.G.	12.1	0.8	6	23	3.9

Medina (1984) calculated reabsorption efficiencies of 48 % and 66 % for N and P, respectively, in Amazonian forests, based on the difference between maximum leaf content and content in newly fallen leaves, so without considering weight loss. If a weight loss of 28 % is assumed these numbers would be 63 and 76 % for N and P. However, if Medina's calculation is based on the highest possible contents during the leaf life span, it does not take into account the likely increase in specific leaf mass after the moment this highest content was reached, implying that the latter numbers are overestimates.

Maximum reabsorption efficiency probably varies between 50 and 60 % for N and between 60 and 70 % for P. Foliar N concentrations generally vary between 10 and 17.5 g/kg and P concentrations between 0.4 and 0.8 g/kg for amazonian forests (Appendix II.1a). Therefore the data in Fig. II.14 were interpreted in two ways as represented by solid line a (lower estimates for the maximum) and solid line b (highest estimates for the maximum).

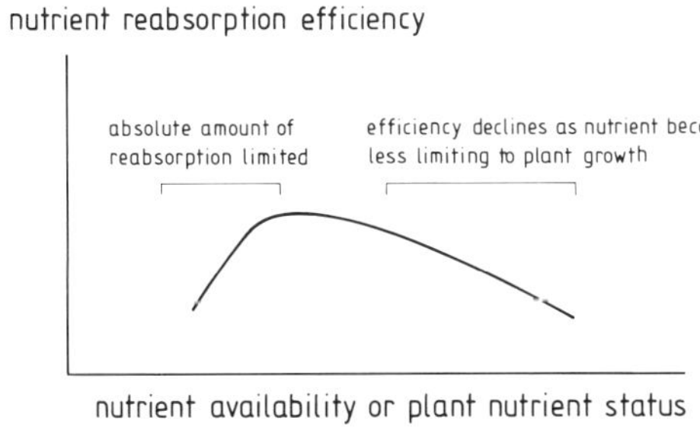

Fig. II.13 A model of the relationship between nutrient availability and nutrient reabsorption efficiency. Source: Lajtha (1987).

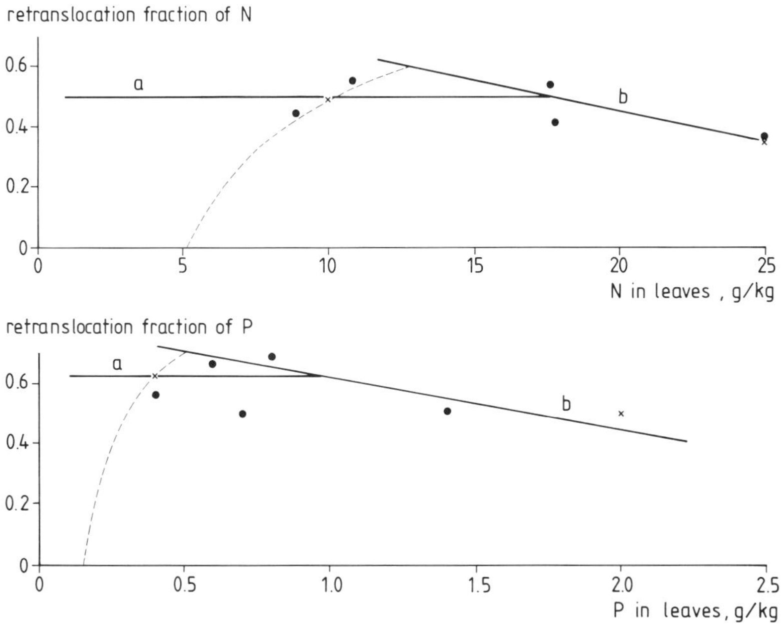

Fig. II.14 Relation between leaf reabsorption efficiency and foliar concentration for N (above) and P (below). Data from Table II.9, ranges calculated in text from data of Vitousek (1982, 1984) are indicated by x. For further comment on dotted and solid lines a and b, see Section 6.2.

6.3 Implications for the model

In the model N and P reabsorption efficiencies are calculated as a function of N and P concentration in the leaves according to the dotted curves and lines b of Fig. II.14:

$$FLRDU_N = MIN\,[(LEAFC_N - 5.1)/LEAFC_N,\ 085\text{-}0.02\ LEAFC_N] \qquad [II.22]$$

$$FLRDU_P = MIN\,[(LEAFC_P - 0.15)/LEAFC_P,\ 0.79\text{-}0.17\ LEAFC_P] \qquad [II.23]$$

where:

$FLRDU_N$ = reduction fraction of N in leaves before leaf fall
$FLRDU_P$ = reduction fraction of P in leaves before leaf fall
$LEAFC_N$ = foliar N concentration, g/kg
$LEAFC_P$ = foliar P concentration, g/kg
MIN [a,b] = the minimum of a and b.

7 SPECIFIC LEAF AREA AND LEAF LIFE SPAN

7.1 Introduction

Mean specific leaf area (SLA) and leaf life span of tropical forests depend on species composition. Evergreen species tend to have low specific leaf area (below 5 m^2 kg^{-1}) and high leaf life span (up to 1.25 year), whereas deciduous species generally have a much higher specific leaf area (10 - 20 m^2 kg^{-1}) and shorter life span (0.25 - 0.75 year) (Medina, 1984; Mooney et al., 1984; Longman & Jenìk, 1987).

Predominance of evergreen species can be seen as an adaptive strategy to nutrient stress, as it increases residence times of nutrients in the organs that determine growth. Thick leaves, i.e. low SLA, are needed under such circumstances to resist herbivory. Deciduous species are more abundant in dry forests than in wet forests. They generally drop their leaves at the beginning of the dry season. This can be seen as an adaptation to moisture stress as a lower leaf area lowers the transpiration. This seasonality in leaf area implies lower residence time of leaf biomass and nutrients and thus less need to resist herbivory. Moreover, as the growing season is shorter, the vegetation has to take maximum advantage of light, water and nutrients during the rainy season. Thinner leaves, i.e. relatively high SLA, may therefore be expected in these dry forests. On the other hand, sclerophylly, i.e. low SLA, may sometimes be an adaptation to very dry climates. Therefore it is hard to predict how the interaction of nutrient and moisture stress works out. Yet, it is probable that SLA is low and leaf life span is high under conditions where nutrients, and not moisture are limiting production. If moisture becomes more limiting, an increasing SLA and decreasing leaf life span may be expected as a result of an increasing proportion of deciduous species. It is assumed that the situation where SLA tends to decrease again as a result of severe moisture stress (sclerophylly) is not relevant for the humid tropical forests we concentrate on.

7.2 Specific leaf area and nutrient status

Figure II.15 that shows there is a positive relation between SLA and N concentration in leaves for tropical forests. A similar relationship was found between SLA and P concentration.

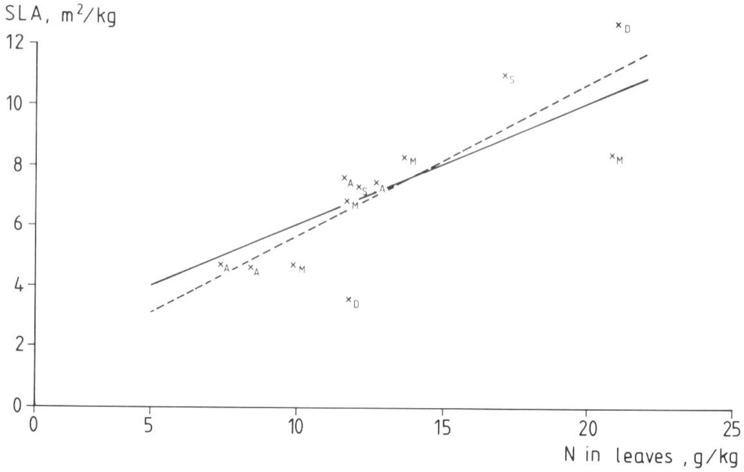

Fig. II.15 Relation between mean specific leaf area of tropical forests and nitrogen concentration of leaves. Data from Medina (1984) and Mooney et al. (1984): x_A = amazonian, x_D = dry, x_M = montane and x_S = semi-deciduous forest. Solid line: linear regression line excluding dry forests from analysis; dotted line: including dry forests.

Although the linear correlation coefficient is higher for N than for P (see Table II.10), it appears that both influence SLA. The lowest SLA value (3.6 m^2/kg) was recorded for an evergreen tropical forest which included sclerophyllous species and which was classified as very dry by Marìn & Medina (1981). The same authors reported the highest value in Fig. II.15 (12.7 m^2/kg) for dry deciduous forest. The relatively low SLA value 8.3 m^2/kg at 21 g/kg of N in leaves for a montane forest may be explained by low P concentration of the leaves. Applying the linear regression equation for P, a SLA of about 6 m^2/kg would be estimated there.

7.3 Leaf life span

The most important factor in leaf fall seasonality is moisture stress, but low light intensities may also play a role (Longman & Jenìk, 1987). As nutrient stress influences species composition in the way described in Section 7.1, long leaf life span may be expected at low nutrient concentrations in the leaves and shorter life span at high nutrient status of the vegetation. Based on the descriptions of Longman & Jenìk (1987) and Medina (1984), a leaf life span range of 0.5 - 1.25 year, depending on nutrient status of the vegetation, seems reasonable in forests with little moisture stress.

7.4 Implications for the model

It is concluded that SLA may be calculated as functions of N and P concentrations in leaves if the forest is not too dry. The linear regression equations that were found excluding dry forests are be used in the model to predict SLA by averaging the estimates derived from N and P concentrations in leaves (Table II.10).

Table II.10 Linear regression equations for relationships between specific leaf area (SLA, m^2/kg) and foliar N concentration ($LEAFC_N$, g/kg) and that between SLA and foliar P concentration ($LEAFC_P$, g/kg)in tropical forests. Data from Medina (1984) and Mooney et al. (1984)

Equation	R^2	Dry forests
$SLA = 0.56 + 0.508\ LEAFC_N$	0.67	included
$SLA = 1.99 + 0.405\ LEAFC_N$	0.64	excluded
$SLA = 2.22 + 6.632\ LEAFC_P$	0.62	included
$SLA = 2.51 + 6.804\ LEAFC_P$	0.50	excluded

For leaf life span it is assumed that the highest nutrient dependent value, being 1.25 years, holds if the concentration of any nutrient is low. Leaf life span is assumed to be 0.5 year if the concentrations of all nutrients are high.

To account for smaller life spans due to moisture and light stress, a treatment analogous to the one in the crop growth model WOFOST is proposed (Van Diepen et al., 1988). It is implicitly assumed here that leaf life span and relative leaf fall rate are each others reciprocal. In WOFOST, both the maximum relative leaf fall rate due to moisture stress and that due to light stress are 0.03 day^{-1}, three times the relative leaf fall rate due to ageing of arable crops like cassava, at 27 °C. Applying the same factor to the leaf fall rate in forests leads to the calculation scheme of Table II.11. Note that the multiplication factors f_{LAI} and f_{MS} do not function until they pass unity. The phenomenon that in case of moisture or light stress oldest leaves are shed first, is in this way indirectly accounted for.

Table II.11 Calculation of relative leaf fall rate of leaves in the model, as a function of LAI, moisture stress and nutrient status

RRLFA	= MINIM ($RRFL_i$) · MAXIMUM ($1, f_{LAI}, f_{MS}$)
$RRLF_i$	= $RRLF_N$ for $RRLF_P$ or $RRLF_K$
$RRLF_N$	= $0.8 < 0.8 + 1.2 \cdot (LEAFC_N - 10)/10 < 2.0$
$RRLF_P$	= $0.8 < 0.8 + 1.2 \cdot (LEAC_P - 0.5) < 2.0$
$RRLF_K$	= $0.8 < 0.8 (LEAFC_K - 5)/10 < 2.0$
f_{LAI}	= $0 < 3 (LAI - LAI_{cr})/LAI_{cr} < 3$
f_{MS}	= $0 < 3 (T_{pot} - T_{act})/T_{pot} < 3$
RRLFA	= relative rate of leaf fall (1/yr)
$RRLF_N$	= relative rate of leaf fall as determined by nitrogen only (yr)
$RRLF_P$	= relative rate of leaf fall as determined by phosphorus only (1/yr)
$RRLF_K$	= relative rate of leaf fall as determined by potassium only (1/yr)
$LEAFC_N$	= nitrogen concentration in leaves (g/kg)
$LEAFC_P$	= phosphorus concentration in leaves (g/kg)
$LEAFC_K$	= potassium concentrations in leaves (g/kg)
f_{LAI}	= multiplication factor for high LAI (light stress)
LAI	= leaf area index (ha(leaf) per ha (ground surface))
LAI_{cr}	= critical LAI, which is about 4.0 in WOFOST
f_{MS}	= multiplication factor for moisture stress
T_{pot}	= potential transpiration in the period considered (mm)
T_{act}	= actual transpiration rate in the period considered (mm)

REFERENCES

Bamber, R.K. (1976). Heartwood, its function and formation. Wood Sci.Technol. **10**: 1-8.

Braakhekke, W.G. (1980). On coexistence: a causal approach to diversity and stability in grassland vegetation. Agricultural Research Reports 902. PUDOC, Wageningen, The Netherlands.

Brown, S. & Lugo, A.E. (1982). The storage and production of organic matter in tropical forests and their role in the global carbon cycle. Biotropica **14**: 161-187.

Brown, S. & Lugo, A.E. (1984). Biomass of tropical forests: a new estimate based on forest volumes. Science **233**: 1290-1293.

Chapin III, F.S. (1980). The mineral nutrition of wild plants. Ann. Rev. Ecol. Syst. **11**: 233-260.

Charley, J.L. & Richards, B.N.(1983). Nutrient allocation in plant communities: mineral cycling in terrestrial ecosystems. in: Lange, O.L., Nobel, P.S. Osmond, C.B. & Ziegler,H. (eds). Ecosystem processes: mineral cycling, productivity and man's influence. Physiological plant ecology **IV**. Springer Verlag, Berlin, Germany.

Coleman, D.C. (1976). A review of root production processes and their influence on soil biota in terrestrial ecosystems. in: Anderson, J.M. & Macfadyen, A. (eds.). The role of terrestrial and aquatic organisms in decomposition processes: 417-434. Blackwell, Oxford, United Kingdom.

De Groof, A. (1988). Toetsing van enige onderdelen van het QUEFTS-model: 1. De relaties tussen aanbod en opname van nutriënten; 2. De relatie tussen de opname van nutriënten en de drogestofopbrengst. (Internal report). Agricultural University, Department of Soil Science and Plant Nutrition, Wageningen, The Netherlands.

De Wit, C.T. (1960). On competition. Verslagen Landbouwkundige Onderzoekingen No. 60.8. PUDOC, Wageningen, The Netherlands.

Edwards, P.J. & Grubb, P.J. (1982). Studies of mineral cycling in a montane rain forest in New Guinea IV: soil characteristics and the division of mineral elements between the vegetation and the soil. Journal of Ecology **70**: 649-666.

Fölster, H., De las Salas, G. & Khanna, P. (1976). A tropical evergreen forest site with perched water table, Magdalena valley, Colombia: biomass and bioelement inventory of primary and secondary vegetation. Acta Oecologica, Oecol. Plant. **11**: 297-320.

Golley, F.B., McGinnes, J.T., Clements, R.G., Child, G.I. & Duever, M.J. (1975). Mineral cycling in a tropical moist forest ecosystem. University of Georgia Press, Athens, USA.

Golley, F.B., Yantko, J.T., Richardson, T. & Klinge, H. (1980a). Biogeochemistry of tropical forests 1: the frequency distribution and mean concentration of selected elements in a forest near Manaus, Brazil. Tropical Ecology **21**: 59-70.

Golley, F.B., Yantko, J. & Jordan, C. (1980b). Biogeochemistry of tropical forests 2: the frequency distribution and mean concentration of selected elements near San Carlos de Río Negro, Venezuela. Tropical Ecology **21**: 71-80.

Greenland, D.J. & Kowal, J.M. (1960). Nutrient content of the moist tropical forest of Ghana. Plant and Soil **12**: 154-174.

Grimm, U. & Fassbender, H.W. (1981). Ciclos bioquímicos en un ecosistema forestal de los Andes Occidentales de Venezuela: inventario de las reservas orgánicas y minerales (N, P, K, Ca, Mg, Mn, Fe, Al, Na). Turrialba **31**: 27-37.

Grubb, P.J. & Edwards, P.J. (1982). Studies of mineral cycling in a montane rain forest in New Guinea III: the distribution of mineral elements in the above-ground material. Journal of Ecology **70**: 623-648.

Hase, H. & Fölster, H. (1982). Bioelement inventory of a tropical (semi-)evergreen seasonal forest on eutrophic alluvial soils, Western Llanos, Venezuela. Acta Oecologica, Oecol. Plant. **3**: 331-346.

Herrera, R., Medina, E., Klinge, H., Jordan, C.F. & Uhl, C. (1984). Nutrient retention mechanisms in tropical forests: the Amazon Caatinga, San Carlos pilot project, Venezuela. in: Di Castri, F., Baker, F.W.G. & Hadley, M. (eds.). Ecology in practice, part I: ecosystem management: 85-97. UNESCO, Paris, France.

Janssen, B.H. (1984). A simple method for the calculation of decomposition and accumulation of 'young' soil organic matter. Plant & Soil **76**: 297-304.

Janssen, B.H. 1986. Een één-parametermodel voor de berekening van de decompositie van organisch materiaal. Vakbl. Biol. **66**: 433-436.

Janssen, B.H., Guiking, F.C.T., Van der Eijk, D., Smaling, E.M.A., Wolf, J. & Van Reuler, H. (1990). A system for quantitative evaluation of the fertility of tropical soils (QUEFTS). Geoderma **46**: 299-318.

Jeník, J. (1978). Roots and root systems of tropical trees: morphological and ecological aspects. in: Tomlinson, P. & M. Zimmerman (eds.). Tropical trees as living systems: 323-349. Cambridge Univ.Press, London.

Jordan, C.F. (1985). Nutrient cycling in tropical forest ecosystems. Wiley & Sons, Chichester, United Kingdom.

Jordan, C.F. & Escalante, G. (1980). Root productivity in an Amazonian rain forest. Ecology **61**: 14-18.

Jordan, C.F. & Herrera, R. (1981). Tropical rain forests: are nutrients really critical? The American Naturalist **117**: 167-180.

Keith, H., Oades, J.M. & Martin, J.K.(1986). The input of carbon to soil from wheat plants. Soil Biol. Biochem. **18**: 445-449.

Klinge, H., Rodriques, W.A., Bruning, E. & Fittkau, J. (1975). Biomass structure in a central Amazonian rain forest. in: Golley, F.B. & Medina, E. (eds). Tropical ecological systems: 115-122. Springer Verlag, New York, USA.

Klinge, H., Farch, K. & Harms, E. (1984). Selected bioelements in bark and wood of native tree species from Central-Amazonian inundation forests. Amazoniana **9**: 105-117.
Klinge, H. & Herrera, R. (1978). Biomass studies in Amazon Caatinga forest in southern Venezuela: standing crop of composite root mass in selected stands. Tropical ecology **19**: 93-110.
Lajtha, K. (1987). Nutrient reabsorption efficiency and the response to phosphorus fertilization in the desert shrub *Larrea tridentata* (DC.). Cov. Biogeochemistry **4**: 265-276.
Longman, K.A. & Jeník, J. (1987). Tropical forest and its environment. 2nd ed. Longman Scientific & Technical, Essex, United Kingdom.
Marín, D. & Medina, E. (1981). Leaf duration, nutrient content and sclerophylly of very dry tropical forest trees. Acta Cient.Ven. **32**: 508-514.
Martin, J.K. & Puckridge, D.W. (1982). Carbon flow through the rhizosphere of wheat crops in south Australia. in: Freney, J.R. & Galbally, I.E. (eds.). Cycling of carbon, nitrogen, sulfur and phosphorus in terrestrial and aquatic ecosystems. Springer Verlag, Berlin, Germany.
Medina, E., (1984). Nutrient balance and physiological processes at the leaf level. in: Medina, E., Mooney, H.A. & Vázquez-Yánes, C. (eds.). Physiological ecology of plants of the wet tropics. Tasks for the Vegetation science **12**: 139-154. Dr. W. Junk Publ., The Hague, The Netherlands.
Mengel, K. & Kirkby, E.A. (1979). Principles of plant nutrition. 2nd ed. International Potash Institute Berne, Switserland.
Mooney, H.A., Field C. & Vázquez-Yánes, C. (1984). Photosynthetic characteristics of wet tropical forest plants. in: Medina, E., Mooney, H.A. & Vàzquez-Yànes, C. (eds.). Physiological ecology of plants in the wet tropics. Tasks for the vegetation science **12**: 113-128. Dr. W. Junk Publ., The Hague, The Netherlands.
Newbould, P. (1968). Methods of estimating root production. in: Functioning of ecosystems at the primary production level: 181-190. UNESCO, Paris, France.
Noij, I.G.A.M. (1988). Een eenvoudig simulatiemodel voor nutriëntenbeperkte fytomassa produktie in tropische bosgebieden: NUTCYC. Verslagen en mededelingen 1988-1. Department of Soil Science and Plant Nutrition, Agricultural University, Wageningen, The Netherlands.
Oades, J.M. (1988). The retention of organic matter in soils. Biogeochemistry **5**: 35-70.
Orman, H.R., & Will, G.M. (1960). The nutrient content of *Pinus radiata* trees. NZ J. Sci. **3**: 510-522.
Ovington, J.D. & Olson, J.S. (1970). Biomass and chemical content of El Verde lower montane rain forest plants. in: Odum, H.T. & Pigeon, R.F. A tropical rain forest, book 3. USAEC, Springfield, USA.
Proctor, J. (1983). Tropical forest litterfall: I problems of data comparison. in: Sutton, S.L., Whitmore, T.C. & Chadwick, A.C. (eds.). Tropical rain forest: ecology and management: 267-273. Blackwell, Oxford, United Kingdom.

Raich, J.W. (1980). Fine roots regrow rapidly after forest felling. Biotropica **12**: 231-232.

Ryan, D.F. & Bormann, F.H. (1982). Nutrient resorption in northern hardwood forests. Bioscience **32**: 29-32.

Rijkelijkhuizen, A. (1987). Nutriëntenvoorziening door varkensdrijfmest. (Internal report). Department of Soil Science and Plant Nutrition, Agricultural University, Wageningen, The Netherlands.

Sanchez, P.A. (1976). Properties and management of soils in the tropics. Wiley. New York, USA.

Sanford, R.L. (1985). Root ecology of mature and successional Amazon forests. (Ph.D. thesis). University of California, Berkeley, USA.

Sauerbeck, D.R. & Johnen, B.G. (1977). Root formation and decomposition during plant growth. in: Soil organic matter studies, vol. I: 141-148. IAEA, Vienna, Austria.

Scheltema, T.G. (1989). Relaties tussen nutriëntenaanbod, nutriëntenopname en drogestofopbrengst. (Internal report). Department of Soil Science and Plant Nutrition, Agricultural University, Wageningen, The Netherlands.

Stark, N. & Spratt, M. (1977). Root biomass and nutrient storage in rain forest oxisols near San Carlos de Río Negro. Tropical ecology **18**: 1-9.

Tanner, E.V.J. (1985). Jamaican montane forests: nutrient capital and cost of growth. Journal of Ecology **73**: 553-568.

UNESCO (1978). Tropical forest ecosystems. UNESCO/UNEP/FAO, Paris, France.

Van Diepen, C.A., Rappoldt, C. Wolf, J. & Van Keulen, H. (1988). CWFS Crop growth simulation model WOFOST (Documentation Version 4.1). Centre for World Food Studies, Amsterdam, The Netherlands.

Van Reuler, H. & Janssen, B.H. (1988). Soil fertility aspects of upland rice in shifting- cultivation systems in south-west Côte d'Ivoir. Proceedings Second International Tropenbos Seminar: 32-36. Tropenbos, Ede, The Netherlands.

Vitousek, P.M. (1982). Nutrient cycling and nutrient use efficiency. The American Naturalist **119**: 553-572.

Vitousek, P.M. (1984). Litterfal, nutrient cycling, and nutrient limitation in tropical forest. Ecology **65**: 285-298.

Vitousek, P.M. & Sanford, R.L. (1986). Nutrient cycling in moist tropical forest. Ann. Rev. Ecol. Syst. **17**: 137-167.

Vogt, K.A., Grier, C.C. & Vogt, D.J. (1986). Production, turnover and nutrient dynamics of above- and belowground detritus of world forests. in: Advances in ecological research **15**: 303-377. Academic Press Inc., London, United Kingdom.

Whittaker, R.M. & Likens, R.H. (1975). The biosphere and man. in: Lieth, H. & R.H. Whittaker, G.E. (eds.). Primary productivity of the biosphere. Ecological Studies 14: 305-328. Springer Verlag, New York, USA.

APPENDICES

APPENDIX II.1 Foliar nutrient concentrations and ratios for various moist tropical forests arranged by soil fertility

After Vitousek & Sanford (1986) unless indicated otherwise (superscript numbers refer to reference list at the end of Appendix II.7).

Site	Concentrations (g/kg)				
	N	P	K	Ca	Mg
Moderately fertile soils					
Panama	-	1.5	15.3	22.9	2.6
Ghana, Kumasi	25.2	1.4	8.5	15.4	4.8
Venezuela[2b]	23.9	1.9	16.9	14.0	4.6
New Britain	20.8	1.5	16.7	20.4	3.0
Zaire, Yangambi	24.5	1.2	19.2	7.0	8.8
Montane sites					
Venezuela-cloud forest	11.7	0.8	5.5	8.7	2.6
-montane forest[3]	16.4	1.1	12.5	4.6	2.4
Puerto Rico-lower montane[4]	11.6	0.7	6.7	5.5	2.0
-elfin forest	9.9	0.6	5.1	6.7	1.6
New Guinea-lower montane[6]	13.2	0.9	7.7	15.0	3.1
Hawaii	6.1	0.8	6.1	7.9	1.8
Jamaica-mull[7]	14.5	0.7	9.3	9.1	3.7
-mor[7]	9.7	0.5	4.2	5.9	2.7
Infertile oxisol/ultisol					
Venezuela	12.7	0.6	4.6	1.9	1.0
Venezuela-San Carlos	17.8	0.6	3.8	1.1	1.1
Brazil	18.4	0.5	5.0	4.2	2.9
Colombia-terrace[11,b]	17.6	0.8	5.2	3.6	1.8
Spodosols/Psamments					
Venezuela-caatinga[e]	11.6	0.7	6.2	4.4	1.5
-caatinga[e]	10.8	0.6	5.8	5.3	3.6
-bana[f]	7.4	0.5	6.4	5.8	1.4
-tall bana[f]	10.3	0.9	6.8	4.6	2.6
-low bana[f]	12.9	1.2	7.2	10.3	2.5
-open bana[f]	8.9	0.4	5.5	6.4	2.2
Brazil-campina[g]	11.1	0.5	6.6	3.7	2.6
-igapó[12,h]	17.3	0.6	6.3	2.5	1.2
Malaysia	8.7	0.2	3.5	7.5	2.0

Site	Ratios			
	N/P	P/K	K/N	ΣKCaMg/N
Moderately fertile soils				
Panama	-	0.098	-	-
Ghana, Kumasi	18.0	0.165	0.34	1.14
Venezuela[2b]	12.6	0.112	0.71	1.49
New Britain	13.9	0.090	0.80	1.93
Zaire, Yangambi	20.4	0.063	0.78	1.43
Montane sites				
Venezuela-cloud forest	14.6	0.146	0.47	1.44
-montane forest[3]	14.9	0.088	0.76	1.19
Puerto Rico-lower montane[4]	16.0	0.105	0.58	1.17
-elfin forest	16.5	0.118	0.52	1.35
New Guinea-lower montane[6]	14.7	0.117	0.58	1.95
Hawaii	7.6	0.131	1.00	2.59
Jamaica-mull[7]	20.7	0.075	0.64	1.52
-mor[7]	19.4	0.119	0.43	1.32
Infertile oxisol/ultisol				
Venezuela	21.2	0.130	0.36	0.59
Venezuela-San Carlos	29.7	0.158	0.21	0.34
Brazil	36.8	0.100	0.27	0.66
Colombia-terrace[11,b]	22.0	0.154	0.29	0.60
Spodosols/Psamments				
Venezuela-caatinga[e]	16.6	0.113	0.53	1.04
-caatinga[e]	18.0	0.103	0.54	1.36
-bana[f]	14.8	0.078	0.86	1.84
-tall bana[f]	11.4	0.132	0.66	1.36
-low bana[f]	10.8	0.167	0.56	1.55
-open bana[f]	22.3	0.073	0.62	1.58
Brazil-campina[g]	22.2	0.076	0.59	1.16
-igapó[12,h]	28.8	0.095	0.36	1.73
Malaysia	43.5	0.057	0.40	1.49

a Average weighted by D^2H-stratum x number of trees.
b Average weighted by proportion of distinghuished vegetation components.
c Overstorey trees only.
d Average value of two forests.
e Sandy soil, occasionally flooded, tall vegetation.
f Sandy soil on higher ground, seasonally high water table, lower-stature vegetation.
g Sandy soil on high ground, low stature vegetation.
h Inundation forest, nutrient-poor black water.
i 18-year old forest.

APPENDIX II.2 Nutrient concentrations (g/kg) in twigs, branches, bark, sapwood and heartwood for various moist tropical forests arranged by soil fertility

After Vitousek & Sanford, (1986). Data are from original authors, as indicated by superscript number (see end of Appendix II.7; for superscript letters see end of Appendix II.1). Dots and arrows indicate the components a particular value refers to. If the boundary of a component was not clearly defined, the arrow has been omitted.

Nutrient	Site code	Twigs	Branches	Bark	Sapwood	Heartwood
N	1.1					
	1.2		-----	---------	3.2-----	-------->
	1.3		-----	---------	4.2-----	-------->
	1.4	8.7[a]	6.7[c]	<--------	3.4[ab]---	-------->
	1.5		-----	---------	2.6-----	-------->
	1.6		-----	---------	2.7-----	-------->
	2.1		3.6	<--------	2.0-----	-------->
	2.2		3.5	<--------	1.6-----	-------->
	2.3	8.2	2.4	5.4	<--------	1.2----->
	2.4	<--------	---------	2.3------	---------	-------->
	2.5	<--------	---------	1.5------	---------	-------->
	3.1a					
	3.2a					
	3.3	7.6	3.7	<--------	4.1------	-------->
	4.1			9.4	<--------	3.1----->

1 Moderately fertile soils.
1.1 Panama[8cd].
1.2 Ghana, Kumasi[1].
1.3 Ghana, Kade[1b].
1.4 Venezuela[2].
1.5 Zaire, Yangambi[1i].
1.6 Zaire[1].

2. Montane.
2.1 Venezuela-montane[3].
2.2 Puerto-Rico-lower montane[4].
2.3 New Guinea-lower montane[6].
2.4 Jamaica-mull[7].
2.5 -mor (peat soil)[7].

3. Infertile oxisols/ultisols.
3.1a Venezuela-San Carlos[9].
3.2a Brazil[10].
3.3 Colombia-terrace[11b].

4. Spodosols/Psamments.
4.1 Brazil-igapó (swamp soil)[12].

Appendix II.2 continued

Nutrient	Site code	Twigs	Branches	Bark	Sapwood	Heartwood
P	1.1			<-------	0.40----	--------
	1.2		-----	--------	0.25----	--------
	1.3		-----	--------	0.26----	--------
	1.4	2.1[a]	1.1[c]	<-------	0.51[ab]--	--------
	1.5		-----	--------	0.5-----	--------
	1.6		-----	--------	0.3-----	--------
	2.1		0.15	<-------	0.12----	--------
	2.2		0.25	<-------	0.07----	--------
	2.3	0.76	0.11	0.25	<--------	0.07----
	2.4	<--------	---------	0.11----	---------	--------
	2.5	<--------	---------	0.11----	---------	--------
	3.1a		0.42	0.20	0.11	0.07
	3.2a		0.23	<--------	0.13----	--------
	3.3	0.28	0.14	<--------	0.12----	--------
	4.1			0.16	<--------	0.18---
K	1.1			<--------	9.0------	-------->
	1.2		-----	---------	3.0------	-------->
	1.3		-----	---------	2.5------	-------->
	1.4	9.9	4.2[c]	<--------	4.0[ab]---	-------->
	1.5		-----	---------	2.7-----	-------->
	1.6		-----	---------	2.5-----	-------->
	2.1		4.8	<--------	3.1-----	-------->
	2.2		1.8	<--------	1.2-----	-------->
	2.3	8.4	2.4	4.5	<--------	1.7---->
	2.4	<--------	---------	2.3------	---------	-------->
	2.5	<--------	---------	1.0------	---------	-------->
	3.1		2.4	2.4	0.85	0.56
	3.2		2.4	<-------	1.4------	-------->
	3.3	2.6	1.7	<-------	1.6------	-------->
	4.1			3.7	<--------	1.8---->

Appendix II.2 continued

Nutrient	Site code	twigs	branches	bark	sapwood	heartwood
Ca	1.1			<--------	11.0-----	------->
	1.2		-----	---------	3.1------	------->
	1.3		-----	---------	5.8------	------->
	1.4	11.0	8.2[c]	<--------	8.0[ab]----	------->
	1.5					
	1.6		-----	---------	5.4------	------->
	2.1		3.3	<--------	1.9------	------->
	2.2		4.3	<--------	2.0------	------->
	2.3	15.7	6.9	22.4	<--------	0.11--->
	2.4	<--------	---------	2.6------	---------	------->
	2.5	<--------	---------	1.4------	---------	------->
	3.1		1.7	2.2	0.26	0.25
	3.2		3.0	<--------	1.8------	------->
	3.3	3.6	2.8	<--------	2.2------	------->
	4.1			4.6	<--------	0.8----->
Mg	1.1			<-------	1.2------	-------->
	1.2		------	--------	1.5------	-------->
	1.3		------	--------	0.94-----	-------->
	1.4	2.1	1.0[c]	<-------	0.65[ab]---	-------->
	1.5					
	1.6		------	--------	0.9------	-------->
	2.1		0.78	<-------	0.54-----	-------->
	2.2		0.85	<-------	0.09-----	-------->
	2.3	0.20	0.6	1.3	<--------	0.5----->
	2.4	<--------	---------	0.5-----	---------	-------->
	2.5	<--------	---------	0.6-----	---------	-------->
	3.1		0.76	0.53	0.17	0.21
	3.2		1.6	<--------	0.99-----	-------->
	3.3	1.0	0.8	<--------	0.59-----	-------->
	4.1			0.52	<--------	0.31---->

APPENDIX II.3 NUTRIENT CONCENTRATIONS (G/KG) IN ROOTS DISTINGUISHED BY DIAMETER CLASSES FOR VARIOUS MOIST TROPICAL FORESTS ARRANGED BY SOIL FERTILITY

After Vitousek & Sanford (1986). Superscript numbers refer to reference list at the end of Appendix II.3, superscript letters to the end of Appendix II.1. Site codes are explained in Appendix II.2, page 96.

Nutrient	Site code	diameter class, mm									
N		1	5/6	13	20	25	30	50/60	100	>100	stumps
	1.1	<-------	13.7----	11.9	<-------	9.1---->	<-------	--------	4.0-----	------->	3.8
	1.3	13.3	8.1	<-------	5.3---->						
	1.4	------	--------	--------	--------	--------	4.7-----	--------	--------	--------	------
	1.5										
	2.1	<-------	--------	--------	6.4-----	--------	--------	------->	<------	2.3-----	------
	2.2	<-------	7.0---->	<-------	--------	--------	--------	5.0-----	-------	--------	------
	2.3	<-------	7.5---->	<-------	--------	5.4-----	------->	4.2	<------	2.7---->	1.4
	2.5	<-------	--------	2.5----	------->	<-------	--------	1.5----	------>	<-------	1.5--
	3.1a										
	3.1b	------	--------	--------	--------	--------	10.5----	--------	-------	--------	------
	3.2a										
	3.2b	<-------	10.0--->	<-------	--------	--------	--------	14.6----	-------	--------	------
	4.2a	<-------	6.7---->								
	4.2b	------	--------	--------	--------	--------	6.3-----	--------	-------	--------	------
	4.3	<-------	10.7--->	<-------	--------	--------	--------	7.4----	-------	--------	------

Appendix II.3 continued

Nutrient	Site code	diameter class, mm									
P		1	5/6	13	20	25	30	50/60	100	>100	stumps
	1.1	------	--------	--------	--------	--------	1.10----	--------	--------	--------	------
	1.3	<-------	0.79--->	0.74	<-------	0.48-->	<-------	--------	0.14----	--------	0.40
	1.4	1.1	1.2	<-------	1.3--->						
	1.5	------	--------	--------	--------	--------	1.10	--------	--------	--------	------
	2.1	<-------	--------	--------	0.38	--------	--------	------->	<-------	0.14----	------
	2.2	<-------	0.3---->	<-------	--------	--------	--------	0.2-----	-------	--------	------
	2.3	<-------	0.32--->	<-------	--------	0.28---	------->	0.19----	<------	0.06--->	0.11
	2.5	<-------	--------	0.1-----	------->	<-------	--------	0.04----	------->	<-------	0.05--
	3.1a	------	--------	--------	--------	--------	0.20----	--------	--------	--------	------
	3.1b	------	--------	--------	--------	--------	0.33----	--------	-------	--------	------
	3.2a	------	--------	--------	--------	--------	0.17----	--------	--------	--------	------
	3.2b	<-------	0.14--->	<-------	--------	--------	--------	0.17----	-------	--------	------
	4.2a	<-------	0.57--->								
	4.2b	------	--------	--------	--------	--------	0.52----	--------	-------	--------	------
	4.3	<-------	0.31--->	<-------	--------	--------	--------	0.07---	-------	--------	------

APPENDIX II.3 continued

Nutrient	Site code	diameter class, mm									
K		1	5/6	13	20	25	30	50/60	100	>100	stumps
	1.1	------	--------	--------	--------	7.3-----	--------	--------	--------	--------	------
	1.3	<-------	5.6---->	5.8	<-------	4.8---->	<-------	--------	0.7-----	------->	1.9
	1.4	5.8	7.1	<-------	6.3---->						
	1.5	--------	--------	--------	--------	6.4-----	--------	--------	--------	-------	------
	2.1	<-------	--------	--------	2.8-----	--------	--------	------->	<------	2.8-----	------
	2.2	<-------	4.2---->	<-------	--------	--------	--------	3.0-----	-------	--------	------
	2.3	<-------	3.9---->	<-------	--------	7.8-----	------->	4.6	<------	4.8---->	3.0
	2.5	<-------	--------	0.1-----	------->	<-------	--------	0.7-----	------>	<-------	1.0----
	3.1a	------	--------	--------	--------	0.67----	--------	--------	--------	--------	------
	3.1b	------	--------	--------	--------	0.83----	--------	--------	-------	--------	------
	3.2a	------	--------	--------	--------	1.7-----	--------	--------	--------	--------	------
	3.2b	<-------	0.69--->	<-------	--------	--------	--------	1.2----	-------	--------	------
	4.2a	<-------	2.9---->								
	4.2b	------	--------	--------	--------	2.5----	--------	--------	-------	--------	------
	4.3	<-------	1.10--->	<-------	--------	--------	--------	0.66---	-------	--------	------

Appendix II.3 continued

Nutrient	Site code	diameter class, mm									
Ca		1	5/6	13	20	25	30	50/60	100	>100	stumps
	1.1	------	--------	--------	--------	18.0----	--------	--------	--------	--------	------
	1.3	<-------	8.8---->	7.8-----	<--------	6.8---->	<-------	--------	3.0-----	------->	3.8----
	1.4	8.9	9.3	<-------	8.1---->						
	1.5										
	2.1	<-------	-------	--------	4.4-----	--------	--------	------->	<------	1.4-----	------
	2.2	<-------	5.6---->	<-------	--------	--------	--------	4.0-----	-------	--------	------
	2.3	<-------	7.1---->	<-------	--------	14.3----	------->	8.2	<------	13.0---->	1.6
	2.5	<-------	--------	11.4----	------->	<-------	--------	1.3-----	------>	<-------	0.6---
	3.1a	------	--------	--------	--------	0.22----	--------	--------	--------	--------	------
	3.1b	------	--------	--------	--------	0.88----	--------	--------	-------	--------	------
	3.2a	------	--------	--------	--------	1.7-----	--------	--------	--------	--------	------
	3.2b	<-------	1.1---->	<-------	--------	--------	--------	2.2-----	-------	--------	------
	4.2a	<-------	1.6---->								
	4.2b	------	--------	--------	--------	1.8-----	--------	--------	-------	------	------
	4.3	<-------	1.2---->	<-------	--------	--------	--------	2.5-----	-------	------	------

Appendix II.3 continued

Nutrient	Site code	diameter class, mm									
Mg		1	5/6	13	20	25	30	50/60	100	>100	stumps
	1.1	------	--------	--------	--------	--------	2.4-----	--------	--------	--------	------
	1.3	<-------	2.0---->	2.5-----	<-------	2.6---->	<-------	--------	0.8-----	------->	0.7
	1.4	2.6	2.7-----	<-------	2.2---->						
	1.5										
	2.1	<-------	--------	--------	1.0-----	--------	--------	------->	<------	4.4-----	------
	2.2	<-------	1.8---->	<-------	--------	--------	--------	0.8-----	-------	--------	------
	2.3	<-------	6.1---->	<-------	--------	1.0-----	------->	1.7	<------	1.3---->	0.8
	2.5	<-------	--------	1.2-----	------->	<-------	--------	0.3-----	------>	<-------	0.8---
	3.1a	------	--------	--------	--------	--------	0.19----	--------	--------	--------	------
	3.1b	------	--------	--------	--------	--------	0.23----	--------	--------	--------	------
	3.2a	------	--------	--------	--------	--------	1.5-----	--------	--------	--------	------
	3.2b	<-------	0.62--->	<-------	--------	--------	--------	0.96----	--------	--------	------
	4.2a	<-------	1.4---->								
	4.2b	------	--------	--------	--------	--------	1.1-----	--------	-------	--------	------
	4.3	<-------	1.5---->	<-------	--------	--------	--------	0.66----	-------	--------	------

References pertaining to Appendices II.1 - II.3

1 Greenland & Kowal, 1960.
2 Hase & Pölster, 1982.
3 Grimm & Fassbender, 1981.
4 Ovington & Olsen, 1970: 10-M radius.
5 Ovington & Olsen, 1970: arithmic means of Table 8.
6 Grubb & Edwards, 1982.
7 Tanner, 1985.
8 Golley et al., 1975.
9 Golley et al., 1980b.
10 Golley et al., 1980a.
11 Fölster et al., 1976.
12 Klinge et al., 1984.
13 Vitousek & Sanford, 1986: Table 8.
14 Vitousek & Sanford, 1986: Table 7.
15 Edwards & Grubb, 1982.

APPENDIX II.4 Nutrient concentrations (g/kg) in chinese cabbage as found in a pot experiment

The treatment codes 0222, etc. refer to the levels of N, P, K, Mg, applied as inorganic fertilizers. The codes om, vm, k, and f refer to different fractions of pig manure. The codes + P K Mg, etc. refer to the nutrients applied as inorganic fertilizer in addition to the pig manure. Data from Rijkelijkhuizen (1987).

Treatment	N	P	K	N/P	P/K	K/N
0222	15.0	4.97	37.2	3.02	0.13	2.48
1111	37.4	4.96	22.9	7.54	0.22	0.61
1131	27.3	5.04	46.9	5.42	0.11	1.72
1133	29.3	4.93	54.8	5.94	0.09	1.87
1311	31.4	6.14	21.4	5.11	0.29	0.68
1313	28.6	6.11	21.5	4.68	0.28	0.75
1331	28.8	6.04	50.2	4.77	0.12	1.74
1333	29.5	6.42	42.3	4.60	0.15	1.43
2220	42.2	6.59	31.4	6.40	0.21	0.74
2222	43.7	6.69	32.9	6.53	0.20	0.75
2224	40.1	6.64	28.8	6.04	0.23	0.72
2242	42.6	6.99	54.6	6.09	0.13	1.28
2422	43.2	6.52	28.5	6.63	0.23	0.66
3111	59.9	6.90	24.6	8.68	0.28	0.41
3113	63.0	7.79	25.2	8.09	0.31	0.40
3131	55.9	6.59	44.6	8.48	0.15	0.80
3133	51.8	5.41	45.7	9.57	0.12	0.88
3311	66.8	9.13	21.0	7.32	0.43	0.31
3313	60.9	8.47	19.8	7.19	0.43	0.33
3331	55.3	7.55	34.6	7.32	0.22	0.63
3333	52.6	6.97	35.2	7.55	0.20	0.67
4222	63.0	7.94	29.3	7.93	0.27	0.47
om + P K Mg	22.4	6.25	48.4	3.58	0.13	2.16
om + N K Mg	50.6	2.47	54.3	20.49	0.05	1.07
om + N P Mg	63.5	8.38	18.0	7.58	0.47	0.28
om + N P K	53.3	8.17	38.0	6.52	0.22	0.71
vm + P K Mg	23.3	5.74	42.0	4.06	0.14	1.80
vm + N K Mg	52.2	1.94	54.7	26.91	0.04	1.05
vm + N P Mg	66.1	10.35	19.1	6.39	0.54	0.29
vm + N P K	54.1	7.30	40.4	7.41	0.18	0.75
k + P K Mg	23.5	7.24	43.4	3.25	0.17	1.85
k + N K Mg	54.8	6.03	49.0	9.09	0.12	0.89
k + N P Mg	64.0	8.80	18.3	7.27	0.48	0.29
k + N P K	53.8	8.24	38.5	6.53	0.21	0.72
f + P K Mg	30.5	6.25	55.2	4.88	0.11	1.81
f[a] + N K Mg	57.1	1.31[a]	41.7	43.59[a]	0.03[a]	0.73
f + N P Mg	62.6	8.76	22.1	7.15	0.40	0.35
f + N P K	53.8	6.65	43.0	8.09	0.15	0.80

[a] Probably salt affected

APPENDIX II.5 Uptake of nutrients by two-months old maize (above-ground + roots) in relation to the supplies

All data in mg/pot. Data derived from De Groof (1988).

Supply			Uptake		
N	P	K	N	P	K
340	7	215	57.1	4.88	63.0
340	7	310	62.3	5.23	83.3
340	7	405	117.1	7.65	156.7
340	7	595	73.1	5.63	106.4
340	7	975	n.a.	n.a.	n.a.
340	23	215	152.8	9.97	135.1
340	23	310	299.7	27.08	320.9
340	23	405	131.0	8.97	209.9
340	23	595	195.1	13.69	309.6
340	22	975	243.0	20.00	482.4
340	39	215	214.2	22.89	149.2
340	39	310	228.7	20.38	238.6
340	39	405	323.4	34.69	386.5
340	39	595	245.8	25.74	368.5
340	39	975	263.5	26.68	528.8
340	72	215	274.6	33.51	171.2
340	72	310	286.2	47.03	230.0
340	72	405	280.6	46.38	318.3
340	72	595	439.9	28.43	503.1
340	72	975	257.3	39.90	596.5
340	138	215	299.3	69.72	173.2
340	138	310	294.5	69.16	272.2
340	138	405	265.7	77.22	294.2
340	138	595	264.1	65.46	414.0
340	138	975	341.0	81.65	791.8
840	7	215	47.1	2.38	50.2
840	7	310	72.4	5.00	78.2
840	7	405	68.3	4.89	75.7
840	7	595	130.0	7.39	150.2
840	7	975	63.3	2.40	106.2
840	23	215	259.1	14.19	163.9
840	23	310	169.8	8.99	204.1
840	23	405	176.3	9.39	224.2
840	23	595	195.6	10.37	259.9
840	23	975	274.3	15.21	409.3

Appendix II.5 continued

Supply			Uptake		
N	P	K	N	P	K
840	39	215	546.5	43.73	172.6
840	39	310	597.6	37.04	265.2
840	39	405	240.8	12.89	252.1
840	39	595	453.8	27.23	447.0
840	39	975	771.8	37.78	953.1
840	72	215	868.3	89.61	232.7
840	72	310	543.9	60.76	232.6
840	72	405	647.9	51.05	267.6
840	72	595	843.9	68.33	613.9
840	72	975	746.8	70.13	801.9
840	138	215	515.8	82.94	137.1
840	138	310	970.7	149.85	356.7
840	138	405	579.5	114.45	250.4
840	138	595	873.6	130.67	621.8
840	138	975	688.3	107.38	840.2
1340	7	215	64.8	4.89	66.5
1340	7	310	n.a.	n.a.	n.a.
1340	7	405	98.1	5.77	107.3
1340	7	595	72.0	4.89	78.1
1340	7	975	70.4	4.57	72.5
1340	23	215	188.5	9.35	139.7
1340	23	310	258.4	12.48	242.6
1340	23	405	192.6	11.39	223.0
1340	23	595	n.a.	n.a.	n.a.
1340	23	975	400.2	17.64	533.4
1340	39	215	535.6	26.22	173.2
1340	39	310	501.8	23.68	251.6
1340	39	405	348.6	53.95	308.4
1340	39	595	413.2	22.62	383.5
1340	39	975	500.6	26.92	535.5
1340	72	215	793.7	51.51	191.1
1340	72	310	826.9	59.48	221.4
1340	72	405	1111.6	82.83	317.1
1340	72	595	1297.1	93.53	666.2
1340	72	975	670.3	45.85	654.7
1340	138	215	928.2	95.66	169.4
1340	138	310	1456.1	134.65	325.1
1340	138	405	1250.7	127.54	371.0
1340	138	595	940.6	79.67	434.9
1340	138	975	1221.0	85.37	821.1

[a] n.a. = not available because these plants were not analysed.

APPENDIX II.6 Uptake of nutrients by two-months old maize (above-ground + roots) in relation to the supplies

All data are in mg per pot. Data derived from Scheltema (1989).

Supply			Uptake		
N	P	K	N	P	K
481	20	303	306.0	21.5	289.7
481	20	468	297.0	21.5	375.7
481	20	634	219.2	15.0	340.3
481	196	303	480.9	160.3	320.6
481	196	468	382.3	146.4	406.7
481	196	634	398.0	144.0	551.2
481	372	303	481.8	225.8	301.1
481	372	468	479.8	183.9	359.8
481	372	634	664.3	263.8	781.6
886	20	303	273.4	16.3	231.8
886	20	468	272.7	14.9	320.4
886	20	634	158.2	8.6	260.9
886	196	303	773.1	163.2	264.4
886	196	468	779.7	135.6	443.6
886	196	634	1274.6[a]	150.1	599.5
886	372	303	868.0	83.1[a]	258.6
886	372	468	916.7	248.9	425.0
886	372	634	906.0	257.4	587.3
1291	20	303	275.0	13.7	244.4
1291	20	468	255.5	14.8	322.8
1291	20	634	262.7	15.1	377.4
1291	196	303	1316.2	198.3	270.4
1291	196	468	1219.3	185.4	423.8
1291	196	634	1282.5	178.5	569.8
1291	372	303	1137.8	264.2	284.4
1291	372	468	1206.0	227.2	393.8
1291	372	634	1420.1	258.9	668.2

a These data were considered unreliable and therefore have not been used in the calculations for Fig. II.1 and Fig. II.2.

PART III

MODEL DESCRIPTION AND SENSITIVITY ANALYSIS

CONTENTS

1 INTRODUCTION

This part of the report describes in details the structure (Section 2) and process formulations (Section 3) of the model DYNAMITE. Background information on the main starting points underlying the model has already been presented in Part I, Section 1 of this report and also in Part II, section 1.2.
The process formulations discussed in the Sections 3.1, 3.2 and 3.3 are derived or modified from NUTCYC, the predecessor of DYNAMITE, and from other models. Process formulations in Section 3.4 are based mainly on relationships developed in Part II.
Section 4 mentions the input data required to run the model. If such data are not available, estimated values have to be introduced.
Section 5 points out how variations in the values of input parameters affect the outcomes of the moisture cycling sub-model.
In Section 6 suggestions are made for modifications and possible extensions of the model.

2 MODEL STRUCTURE

2.1 Discretization in time, depth and space

For numerical analysis of the behaviour of a soil-vegetation system in time we have to discreticize the system. Discretization may lead to response artifacts. A well known example is the numerical dispersion resulting from subdividing a soil column in discrete layers. This artifact, however, can be used to simulate physical dispersion resulting from molecular diffusion and hydrodynamic dispersion. Discretization in time will cause, amongst other effects, smoothing of irregularly distributed boundary fluxes like rainfall, temperature, erosion and atmospheric deposition of dust. The effect of discretization can be drastic and therefore should be included in a sensitivity analysis. Besides allowing numerical analysis, discretization in time and space may also be imposed by the level of detail by which input data are available. Especially with respect to studies in tropical regions, the time and depth grid in which model parameters may be collected often will be rather coarse.

The present model uses a fixed time step that may vary from two weeks to one year. Smaller time steps are prefered when conditions for nutrient and moisture cycling are limiting forest development during specific, short-lasting periods within the year. The length of the time step will also depend on the purpose of the simulation experiment. For simulation of timber felling or burning a shorter time step has to be chosen than for the simulation of a long-term steady state in a tropical forest.

The present model distinguishes one soil layer for nutrient uptake and two layers for water uptake. With respect to tropical forests on slopes, lateral transport of water, nutrients and soil material is very important, and lateral inputs from surface and subsurface runoff and erosion have to be considered. In the present model only vertical transport is considered, which limits the applicability. However, it is possible to modify the model for simulation of nutrient and moisture cycling in a sequence of soil profiles along a hill slope, similar to the concept of sub-catchments in the watershed acidification model ILWAS (Goldstein et al., 1984). Subdivision of the system in more soil layers and the distinction of soil segments along the toposequence, will strongly increase simulation time and therefore of the model.

2.2 Chemical constituents and element pools

Before describing the model formulations in detail, an overview is given of the chemical constituents, of the pools in which the constituents can occur and of the processes that can change these pools.

In the model the following constituents are distinguished:

- phosphorus
- nitrogen
- potassium
- carbon

and directly related to carbon:

- dry matter

Four vegetation components are distinguished:

- leaf
- wood
- fine roots
- coarse roots

The primary organic pools, which are directly filled by vegetation die-back are:

- leaf litter (forest floor leaves)
- wood litter (forest floor wood)
- fine root debris
- coarse root debris.

The following pools of soil inorganic matter are distinguished:

- stable organic matter
- moderately labile organic matter
- labile organic matter

The following pools of soil inorganic matter are distinguished:

- inert organic matter
- stable inorganic matter
- labile inorganic matter
- adsorbed inorganic matter

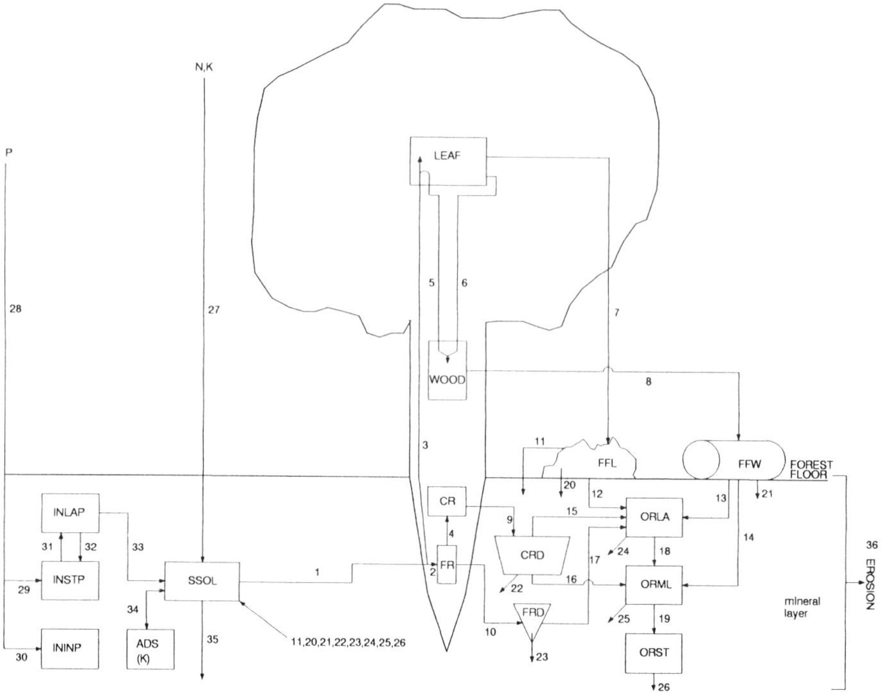

N,K
P
LEAF
5
6
28
27
7
WOOD
8
3
11
FFL
FFW
FOREST FLOOR
20
12
13
21
INLAP
CR
15
ORLA
9
4
14
36
31
32
33
CRD
24
18
EROSION
INSTP
SSOL
1
2
FR
17
29
22
16
ORML
mineral layer
34
FRD
25
19
10
ININP
ADS (K)
35
11,20,21,22,23,24,25,26
23
ORST
30
26

Fig. III.1 Schematic representation of the processes incorporated in the model DYNAMITE. The numbers refer to:

1	plant nutrient uptake from soil solution (SSOL)
2	nutrient uptake in fine roots (FR)
3	nutrient transport from fine roots to leaves (LEAF)
4	nutrient transition from fine to coarse roots (CR)
5	nutrient overflow from leaves to wood
6	nutrient retranslocation from leaves to wood
7	leaf fall
8	wood fall
9	coarse-root dying
10	fine-root sloughing
11	K leaching from forest-floor leaves (FFL)
12	nutrient transfer from forest-floor leaves to organic labile pool (ORLA)
13	nutrient transfer from forest-floor wood (FFW) to organic labile pool
14	nutrient transfer from forest-floor wood to organic moderately labile pool (ORML)
15	nutrient transfer from coarse-root debris (CRD) to organic labile pool
16	nutrient transfer from coarse-root debris to organic moderately labile pool
17	nutrient transfer from fine-root debris (FRD) to organic labile pool
18	nutrient transfer from organic labile to organic moderately labile pool
19	nutrient transfer from organic moderately labile to organic stable pool (ORST)
20	mineralization of forest floor leaves
21	mineralization of forest floor wood
22	mineralization of coarse-root debris
23	mineralization of fine-root debris
24	mineralization of organic labile pool
25	mineralization of organic moderately labile pool
26	mineralization of organic stable pool
27	wet deposition of N and K
28	total dry deposition of P
29	dry deposition of P to inorganic stable phosphorus (INSTP)
30	dry deposition of P to inorganic inert phosphorus (ININP)
31	transfer of P from inorganic stable to inorganic labile phosphorus (INLAP)
32	transfer of P from inorganic labile to inorganic stable phosphorus
33	transfer of P from inorganic labile pool to soil solution
34	desorption and adsorption of K
35	leaching of nutrients from soil solution (SSOL)
36	erosion of organic and inorganic nutrient pools

2.3 Process diagram and sequence of calculations

In Figure III.1 a schematic view is given of the relationships between the various constituent pools and of the processes in a tropical forest ecosystem, as they are incorporated in the DYNAMITE model.

The sequence of calculations before the start of the simulation is:

1) Reading and echoing the input data. Initialization of output files.
2) Initialization of initial contents of organic element pools. Calculation of dissimilation constants.

The sequence of calculations during the simulations is:

1) adjustment of three inorganic pools (inert, stable, labile) and the seven organic pools; integration of major incoming, outgoing and internal element fluxes; calculation of total element pools in soil, vegetation and forest floor;
2) calculation of element ratios in organic pools;
3) simulation of hydrology; calculation of transpiration, percolation and water contents;
4) calculation of erosion fluxes;
5) calculation of atmospheric deposition fluxes;
6) calculation of mineral weathering fluxes;
7) calculation of dissimilation fluxes from and transfer fluxes between organic pools;
8) separate calculation of dissimilation and transfer fluxes from the leaf litter pool, which in contrast to other organic pools is divided into sub-pools of different stability;
9) calculation of nutrient uptake by the plant, and growth of the different vegetation components;
10) solution of the overall mass balance equation, including linear adsorption. Calculation of the leaching fluxes;
11) output of results to files.

3 PROCESS FORMULATIONS

The DYNAMITE model can be divided into a Moisture Cycling Model (or Water Balance Model) and a Nutrient Cycling Model. The discussion of the latter is split into sections on inorganic pools and fluxes, organic pools and fluxes, nutrient uptake and vegetation growth, litter fall and root turnover. The two submodels are brought together in the section on chemical balance.

3.1 Moisture Cycling Model

3.1.1 Main characteristics of the used model

3.1.1.1 General description

A moisture cycling model calculates evaporation and transpiration, soil water content and soil water flux. In the calculation of evaporation and transpiration, interception plays an important role. The interception sub-model used in DYNAMITE had to be calibrated first, and this is discussed in Section 3.1.1.2.

There are two major types of models for water flows in the soil. On the one hand, there are deterministic models based on the Darcy flow equation and the continuity equation, where water transport is driven by the depth gradient of the moisture potential. On the other hand, there are empirical box models where flows to and from a box are determined by the water content. Darcy models require detailed information with respect to hydraulic properties. The measurement of hydraulic properties is tedious. The application of hydraulic functions derived from small columns in the laboratory to the field is a matter of debate. Box models require only some characteristic water contents and a maximum percolation rate, and therefore appear to be more suitable for application to tropical regions, where there is limited opportunity to collect hydraulic data. We adapted a box model as used in the ILWAS model (Goldstein & Chen, 1983), to incorporate capillary rise, which may be an important source of water in tropical regions.

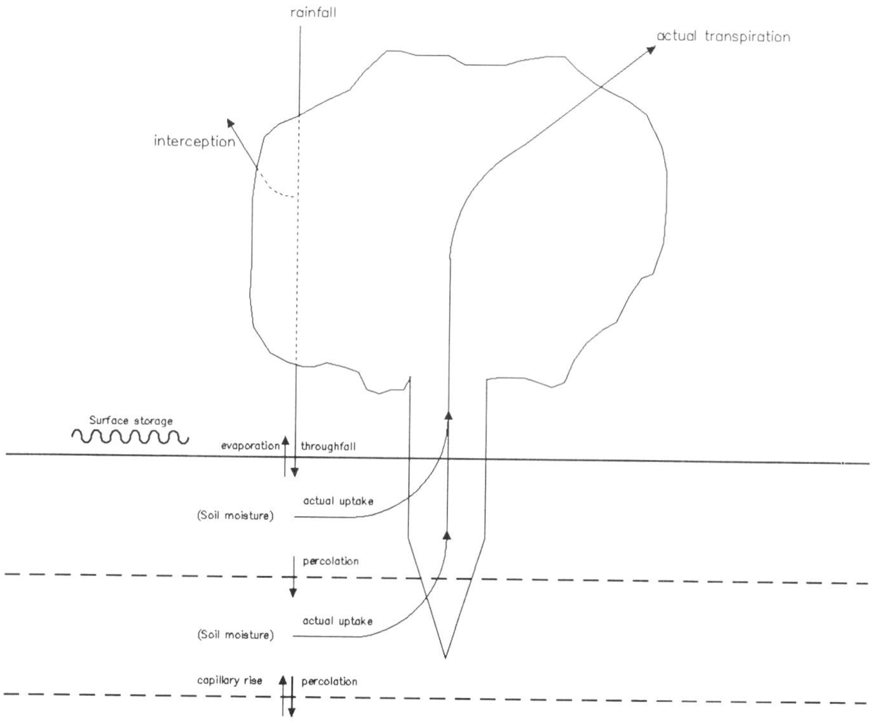

Fig. III.2 Schematic representation of the water balance sub-model in DYNAMITE.

3.1.1.2 Calibration of the interception sub-model

In the presence of a vegetation only part of the precipitation will infiltrate into the soil due to direct interception by leaves and evaporation from the leaves (Fig. III.2). An excellent review of the interception process and interception models is given by Van Roestel (1984)to which one is refered to.
In our study, the empirical relationships by Jackson (1975) were applied to scarce monthly interception data of the Taï region, Ivory Coast (Collinet et al., 1984):

$$I_a = a + b \cdot P \qquad [III.1]$$

$$I_a = a + b \cdot P + c \cdot P^2 \qquad [III.2]$$

$$I_a = a + b \cdot \ln P \qquad [III.3]$$

where I_a is Interception and P is Precipitation.

All models were roughly equally suitable to fit the annual interception but differences between predicted monthly values and observations of 30-50% are common (Fig. III.3). These differences are caused by the variation of precipitation intensity and evaporative demand per event which are not taken into account. The square of the correlation coefficient between observed and predicted monthly interception for all three Jackson's models is about 0.65.

The empirical interception model by Bultot et al. (1976) includes the effects of rainfall intensity and evaporative demand. We modified his equation a little, yielding:

$$I_a = (a \cdot P^2 + b \cdot P)\,(2Q_{av}/(Q + Q_{av}))c \cdot E_{pan} \cdot (1\text{-}e^{0.4.LAI}) \qquad [III.4]$$

where:

- P = precipitation (mm/day)
- Q = average precipitation intensity on days with rain during the time step in the model (mm/day)
- Q_{av} = average precipitation intensity on days with rain during the whole simulation period (= > time step) (mm/day)
- E_{pan} = pan evaporation (mm/day)
- LAI = leaf area index (ha/ha)
- e = base of natural logarithm (2.718)
- a,b,c = regression coefficients

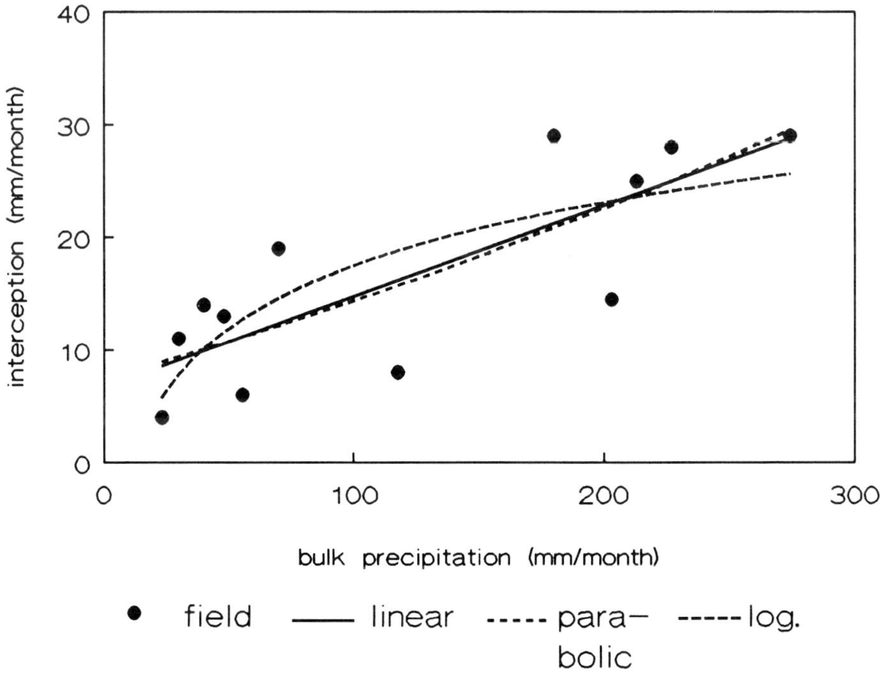

Fig. III.3 Comparison of three empirical interception models and observed (Collinet et al., 1984) monthly interception for the Taï forest, from May 1978 to April 1979.

The model by Bultot et al. originally uses daily time steps. Before use, the model should be calibrated to experimental data for larger time steps. The model can be seen as a refinement of Jackson's parabolic model (Equation III.2 with a=0, Fig. III.4). The correction term for precipitation intensity, $2Q_{av}/(Q + Q_{av})$ tells that intenception will be enhanced if $Q <_{av}$, and interception will be reduced, if $Q > Q_{av}$.

The correction factor for E_{pan} is c. For $E_{pan,av}$, which is the average E_{pan}, $c \cdot E_{pan}$ should be 1, so $c = 1/E_{pan,av}$. Predicted I_a should approach P for increasing E_{pan} and decreasing Q. However, in the Bultot model there is no upper limit for I_a.
In Equation III.4, interception is also corrected for variation in LAI, by multiplying with the factor 1-exp (0.4 LAI), in accordance with Equation III.10 for soil evaporation. The correction for LAI was necessary to reduce interception after (partial) clearcutting.

The values for a en b can be obtained by fitting the model to field data of I_a, P, Q and E_{pan}. If such data are not available, two realistic combinations of I_a, P, Q and E_{pan} may be estimated in order to provide two equations for solution of a and b. The choice of the combinations is then somewhat arbitrary. The shape of the resulting relationship between I_a and P should be similar to that in Fig. III.4.

In this study one set of data was available from Taï, Côte d'Ivoire (Table III.2). They are used to illustrate the procedure. The months January and September are taken as examples of a dry and a wet situation, respectively. The value of Q_{av} is found as ΣRR/ΣDAYWR, being 1832/134 = 13.67. The value of LAI is assumed to be 6 ha/ha, so (1 - exp (0.4 LAI)) is 0.9093. Collinet et al. (1984) found that interception varies between 25% (dry periods) and 10% (wet periods). Substition of these values and the relevant data from Table III.2 in Equation III.4 results in the following equations for the months January and September, respectively

$$0.25 \cdot 21 = (a \cdot 212 + b \cdot 21)\,(2 \cdot 13.67/(21/14 + 13.67)(110/104.5) \cdot 0.9093 \qquad [III.5]$$

$$0.10 \cdot 293 = (a \cdot 293^2 + b \cdot 293)\,(2 \cdot 13.67/(293/12 + 13.67)(82/104.5) \cdot 0.9093 \qquad [III.6]$$

Equations III.5 and III.6 yield: $a = 5.3214\ 10^{-5}$, and $b = 0.17959$
It was not possible to validate the Bultot model for the Taï forest. The data set from Collinet et al. (1984), used to fit Jackson's models, did not include pan evaporation and rainfall intensity, which are required for application of Bultot's model. Bultot's model was provisionally validated using another data set for Taï forest (Fig. III.5) including rainfall intensity and pan evaporation but lacking interception observations (Casenave et al., 1980). The discrepancy between interception predicted by the parabolic model and observed interception (Fig. III4) is of similar magnitude

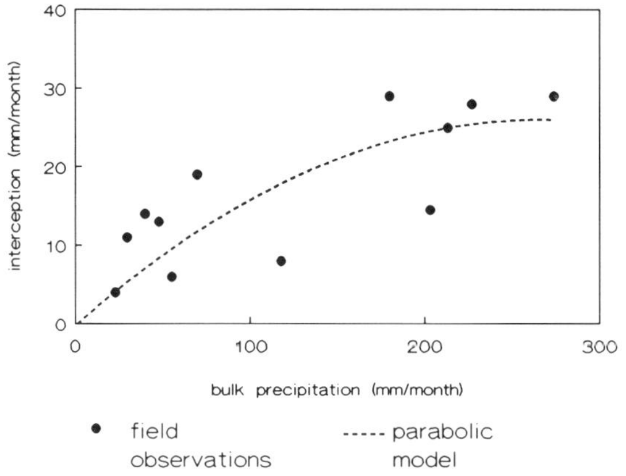

Fig. III.4 Comparison of parabolic interception model without zero degree term and observed (Collinet et al., 1984) monthly interception for the Taï forest, from May 1978 to April 1979.

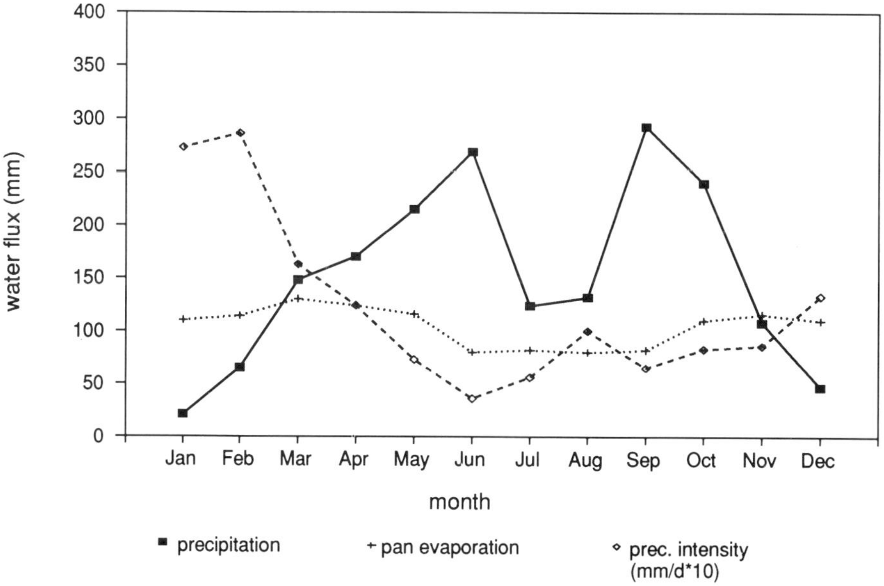

Fig. III.5 Observed rainfall, pan evaporation and rainfall intensity for calculation in Fig. III.4a. (Casenave et al., 1980).

as the discrepancy between predicted interception by the parabolic model and Bultot's model (Fig. III.6). This similarity qualitatively supports the corrections made in Bultot's interception model for variation of monthly evaporation and average rainfall intensity.

3.1.2 Flow scheme and sequence of calculations

Like others, the present Moisture Cycling Model calculates the actual transpiration, soil water content and soil water flux. Groundwater is not yet considered. The dynamic effect of the nutrient cycling model on the moisture cycling model results from the relationship between potential transpiration and leaf area index (LAI). Dynamic effects of the moisture cycling model on the nutrient cycling model are various: the soil water flux determines the solute flux, the transpiration rate may limit the vegetation growth and hence affect the nutrient distribution in the plants, and the soil water content can limit nutrient uptake. The time steps and soil-layer thicknesses for the moisture cycling and nutrient cycling model are not necessarily the same.

The sequence of calculations (Fig. III.2) for the Water Balance Model is:

1) reading of precipitation and potential evapotranspiration, number and thickness of soil layers, soil water characteristics, initial water contents for every soil layer and water uptake distribution with depth;
2) copying of the LAI (total leaf area/land area) calculated in the nutrient cycling model;
3) calculation of the interception losses and correction of the evaporative demand from soil and vegetation;
4) distribution of corrected evapotranspiration over potential soil evaporation and transpiration according to the LAI;
5) calculation of the soil water content below which reduction of water uptake by roots will start;
6) calculation of actual soil evaporation;
7) calculation of infiltration and ponding;
8) calculation of the new water content, water uptake by the vegetation and soil water flux;
9) calculation of "back flow" of water if the soil compartment is saturated and the infiltration is higher than the maximum percolation. This surplus of water will infiltrate in the next time step;
10) averaging and summation of hydrologic parameters for the nutrient cycling model.

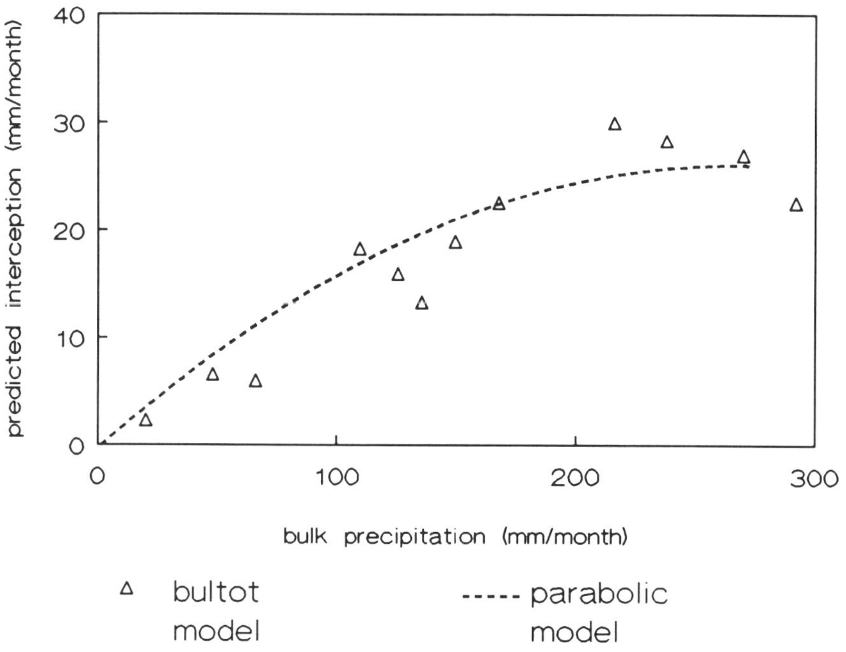

Fig. III.6 Predicted interception as a function of precipitation by the parabolic model and the Bultot model.

3.1.3 The water balance equation

A general formulation of the water balance per time step is:

$$\Theta_2 = (\Theta_1 \cdot D + (J_{in} - J_{out} - E_{act} - W_{act})/D \qquad [III.7]$$

where:

Θ_1	=	water volume fraction at the beginning of the time step (mm^3/mm^3)
Θ_2	=	water volume fraction at the end of the time step (mm^3/mm^3)
D	=	thickness of soil layer (mm)
J_{in}	=	incoming and outgoing water flux (mm)
E_{act}	=	actual soil evaporation (mm)
J_{out}	=	outflow water flux (mm)
W_{act}	=	actual water uptake flux (mm)

Equation III.7 is solved implicitly in time for every soil layer (See Section 3.1.7). J_{in} is a boundary condition. Both the soil water flux and the actual water uptake by roots (sink) are continuous functions of the soil water content. The flux and sink functions are characterised only by the, soil specific, characteristic water volume fractions at saturation (pF =0), field capacity (pF =2) and wilting point (pF = 4.2). In general, data on these characteristic water contents are easily available, which makes the water balance model more easily applicable than models based on the Darcy equation. The soil water flux and water uptake functions in Equation III.7 are solved for the mean of the soil water volume fraction at the end of the previous (Θ_1) and the present time step (Θ_2), the latter being unknown. An implicit solution method for the new water content allows the use of larger time steps than an explicit solution method. The use of large time steps may be necessary when input data are scarce, or to reduce computation time.

3.1.4 Precipitation, evapotranspiration, interception

Precipitation
Precipitation is an input parameter. In mosts cases precipitation data will be available on a daily basis. Simulation of hydrology with time steps longer than one day will smooth water availability, and will usually result in a more efficient water use by the vegetation than simulation with time steps of one day.
Because of interception by the vegetation only part of the precipitation will reach the soil. The water input flux after passage of the canopy is called throughfall.

Potential evapotranspiration
The basic input variable for calculation of evapotranspiration is the open water evaporation E_0 or Pan evaporation (E_{pan}), which may be known for the specific location or can be calculated by empirical or deterministic models (Thornthwaite and Holzman, 1939; Penman, 1948). E_0 or E_{pan} is transformed to the sum of evaporative demand for vegetation and soil ($ET_{pot'}$ in mm/yr) by means of an empirical crop factor:

$$ET_{pot} = f_1 \cdot E_{pan} \qquad [III.8]$$

Next, the $ET_{pot'}$ will be reduced for the part of the evaporative demand which is satisfied by direct evaporation from the canopy (E_i). This amount of water may be measured as the difference between precipitation and throughfall or may be estimated from an interception model (Section 3.1.1.2). The reduction is generally less than proportional (Singh and Sceisz, 1979), because part of the energy required for E_i is obtained from stored sensible heat in the forest and heat advection from the surrounding area.

$$ET_{pot} = ET_{pot'} - f_2 \cdot E_i \qquad [III.9]$$

with $0 < f_2 < 1$

Next, ET_{pot} is distributed over potential soil evaporation (E_{pot}) and potential transpiration (T_{pot}) according to the Leaf Area Index (LAI) (Driessen, 1986):

$$E_{pot} = ET_{pot} \cdot \exp(-0.4 \cdot LAI) \qquad [III.10]$$

$$T_{pot} = ET_{pot} - E_{pot} \qquad [III.11]$$

See Section 3.4.3.2 for calculation of LAI.

Interception

In Section 3.1.1.2 it is explained that interception is calculated as a function of rainfall, rainfall intensity during the time step and during the whole simulation period, pan evaporation and leaf area index (Equation III.4).

3.1.5 Infiltration, percolation and capillary rise

As indicated in Section 3.1.1.1, a box model incorporating capillary rise is used for the water flux in the soil. Capillary rise is the upward flow from a saturated or from a wet but still unsaturated soil compartment.

The water flow rate (J_θ) is calculated per time step by:

$$J_\theta = J_{cr} \frac{e^{a\theta} - b}{b - 1} \qquad [III.12]$$

where:

J_{cr} = maximum capillary rise (mm)

b = $e^{a\theta_{fc}}$

Θ_{fc} = water volume fraction at field capacity

a = constant

Values of J_{cr} and a depend on soil texture and organic matter content. Some typical values are given in Table III.1. Equation III.12 predicts a negative upward flow (capillary rise) for $\Theta < \Theta_{fc}$ and a positive downward flow (drainage) for $\Theta > \Theta_{fc}$. There is no flow when $\Theta = \Theta_{fc}$ (Fig. III.7). Water flows from or to a soil layer, predicted by Eq. III.12, are a continuous function of Θ in this soil layer only. In other words, Eq. III.12 assumes that both drainage from and capillary rise to a soil layer are not limited by Θ in the underlying layer. The value of a is obtained by non-linear optimization, using estimates of the maximum rates of capillary rise and percolation for input. If there is evidence that this assumption leads to large errors of predicted water fluxes, Eq. III.12 may be multiplied with a term, α, including Θ in the underlying layer, e.g. of the type:

$$\alpha = (\Theta/\Theta_s)^\beta \qquad [III.13]$$

where:

Θ_s = Θ at saturation for underlying layer

β = sensitivity (0.01-0.1)

The value of α will be close to 1, except near $\Theta = 0$, where α will be forced to a value of 0, thus blocking capillary rise in case the underlying layer is too dry.

If the throughfall flux in a time step is larger than J_{cr} of the surface layer, surface storage (or ponding) will occur. This amount of water becomes available for infiltration in the next time step.

3.1.6 Water uptake by vegetation and soil evaporation

The transpirative demand (Equation III.11) is distributed over the soil layers according to the root water uptake distribution:

$$W_{pot.j} = T_{pot} \cdot WUF_j \qquad [III.14]$$

where:

$W_{pot,j}$ = potential water uptake by roots in soil layer j (mm/yr)

WUF_j = water uptake fraction for roots in soil layer j (mm/yr)

As a first approximation, WUF may set equal to the distribution of fine roots. The actual uptake of water by roots from a soil layer, ($W_{act,j}$) is calculated by:

$$W_{act.j} = S \cdot W_{pot.j} \qquad [III.15]$$

where: S = Sink term; $0 \leq S \leq 1$

S is a continuous function of Θ :

$$S = \frac{1}{1 + a \cdot e^{-b(\theta s - \theta)}} \qquad \frac{1}{1 + a \cdot e^{-c\theta}} \qquad [III.16]$$

where: a, b, c = constants

S is a bell-shaped curve (Fig. III.8), which is forced through four points: $S_{\theta_s} = 0$, $S_{(\theta_s - d)} = 1$, $S_{\theta_r} = 1$, $S_{\theta_{wp}} = 0$. Θ_s is Θ at saturation, Θ_r is Θ below which water uptake by roots is reduced due to drought, Θ_{wp} is Θ at wilting point. The d is minimum volumetric soil gas fraction required for optimum res-

Table III.1 Values for maximum capillary rise (J_{cr}) and the constant a of Equation III.12.

Horizon	Texture	J_{cr} (mm/day)	a
surface	sand	0.54	42.8
	loamy sand	1.84	44.5
	clay loam	0.43	69.4
subsoil	sand	0.29	45.7
	loamy sand	2.02	51.7
	clay loam	0.73	66.2

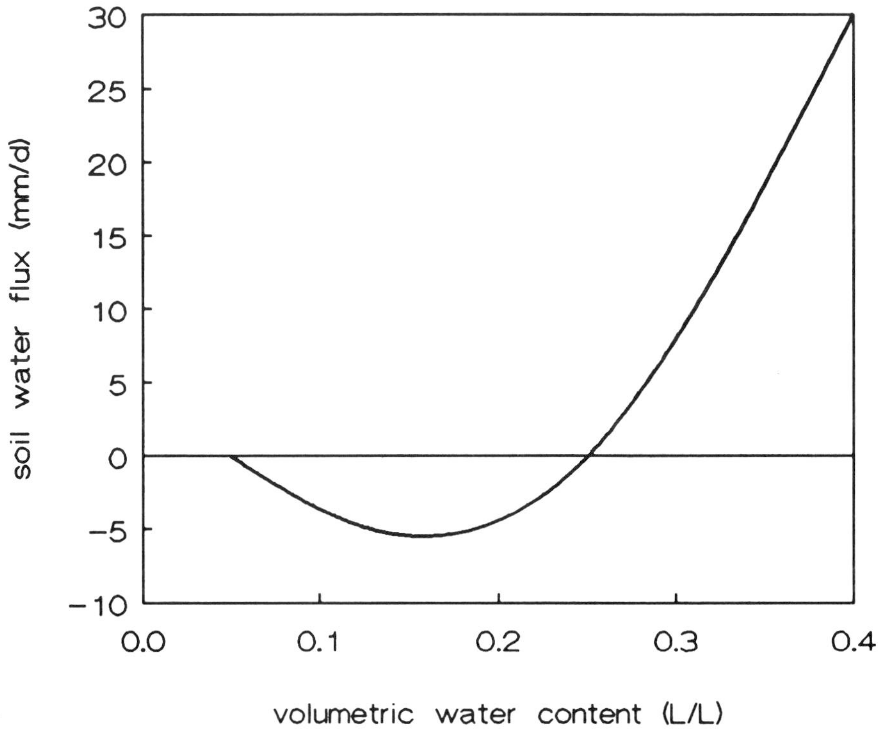

Fig. III.7 Relationship between soil water flux and volume fraction of water for the Taï forest used in DYNAMITE. Negative values indicate upward fluxes or capillary rise.

piration of roots. The values of a, b, c are obtained by non-linear fitting. In accordance with the Sink Term concept of Feddes (1978), Θ_r is a function of Θ_{fc} and the transpiration demand, T_{pot} .

$$\Theta_r = \Theta_{wp} + (1 - \sigma) \cdot (\Theta_{fc} - \Theta_{wp}) \qquad [III.17]$$

The parameter σ is discontinuous function of the transpirative demand (Driessen, 1986). For forests the following functions were selected.

$$\begin{aligned} \sigma &= 0.30 && \text{for} && T_{pot} > 10 \text{ (mm/d)} \\ \sigma &= 3/\text{Tpot} && \text{for} && 6 < T_{pot} < 10 \\ \sigma &= 0.9 - 0.1\,(T_{pot} - 2) && \text{for} && 2 < T_{pot} < 6 \\ \sigma &= 0.9 && \text{for} && T_{pot} < 2 \end{aligned} \qquad [III.18]$$

As Θ_r is a function of T_{pot}, the parameters of Eq. II.15 have to be modified during the simulation each time T_{pot} changes.

The actual transpiration is the sum of actual water uptake for all depths:

$$T_{act} = \Sigma_{j=1}^{n} W_{act.j} \qquad [III.19]$$

The actual soil evaporation (E_{act}) is a function of Θ in the surface layer:

$$E_{act} = E_{pot} \cdot (\Theta - \Theta_{ad})/(\Theta_{fc} - \Theta_{ad}) \qquad [III.20]$$

with Θ_{ad} = the water content after air drying

3.1.7 Solution of water balance

For solution of the new values of Θ , the equation for Jin, Jout, and Wact are substituted in the relaxed water balance equation (Eq. III.7):

$$(\Theta_1 - \Theta_2)D + J_{in} - J_{out\,\theta_2} - W_{act\,\theta_2} = 0 \qquad [III.21]$$

from which Θ_2 is solved numerically by the Newton Raphson technique (Press et al., 1986).

For the surface layer, J_{in} is equal to:

$$J_{in} = P - I_a - E_{act} \qquad [III.22]$$

For the subsurface layers J_{in} is equal to J_{out} from the overlying layer.

J_{out} is substituted by Eq. III.12 and W_{act} is substituted by Eq. III.14. Both J_{out} and W_{act} are a function of unknown Θ_2. As the formulations of J_{out} and W_{act} are highly non-linear, and because the change of Θ within one time step may be considerable, the argument (Θ') of J_{out} and W_{act} is not Θ_2 but the arithmetical mean of Θ_1 and Θ_2:

$$\Theta' = 0.5\,(\Theta_1 + \Theta_2) \qquad [III.23]$$

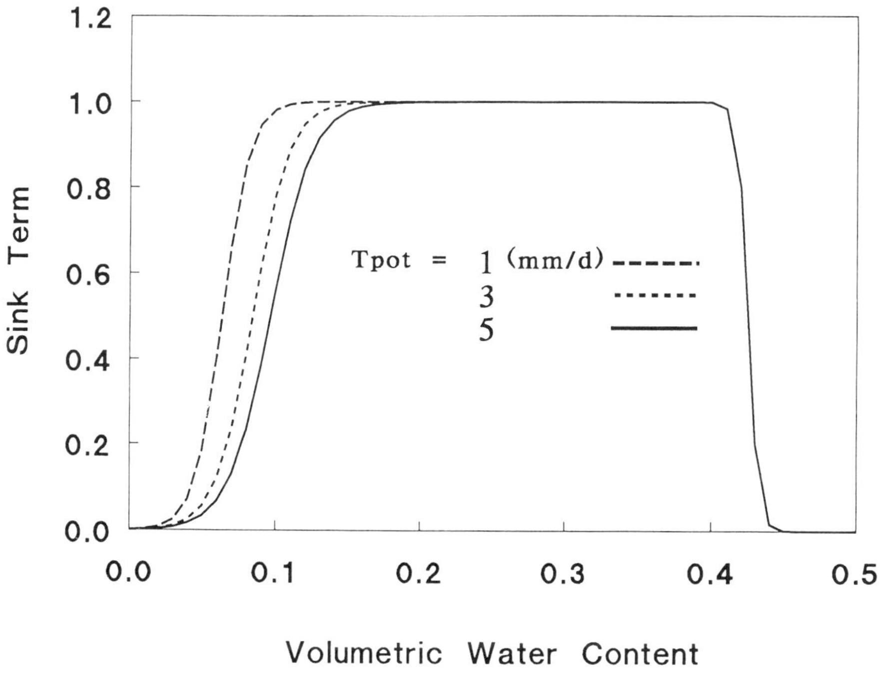

Fig. III.8 Relationship between the ratio of actual over potential root water uptake (sink term) and volume fraction of water for three values of T_{pot}.

3.2 Inorganic pools and fluxes

3.2.1 Atmospheric deposition

Atmospheric deposition is the sum of wet and dry atmospheric deposition. Wet deposition is direct element input via precipitation. Sea spray may contribute substantially to wet deposition depending on the distance from the sea, and prevailing wind directions. Dry deposition is atmospheric input of gases and particles to soil and canopy. Deposition of dust from arid regions or volcanoes may be an important source of phosphorus and potassium. Contrary to wet deposition and dry deposition of gases, deposition of dust generally supplies elements in a form poorly available to plants. In the model only total atmospheric deposition is considered, which is assumed to be constant within the year. Hence:

$$RDEP_i = ARDEP_i/n \qquad [III.24]$$

where:

$RDEP_i$ = rate of deposition of substance i (kg/ha per time step)
$ARDEP_i$ = annual rate of deposition of i (kg/ha per year)
n = number of time steps per year

Atmospheric deposition is considered for phosphorus, nitrogen and potassium (Fig. III.1: Flow 27 and 28).

3.2.2 Erosion

Erosion is considered for all organic and inorganic element pools. The rate of erosion is modeled as a first order process. The rate constant may vary for different pools:

$$REROS_k = POOL_k \cdot FER_k \qquad [III.25]$$

$$FER_k = 1 - \exp((-RREROS_k) \cdot \partial t)$$

where:

$REROS_k$ = rate of erosion for pool $_k$ (kg ha^{-1} per year)
$POOL_k$ = size of pool k (kg/ha)
FER_k = erosion fraction of pool k per time step
$RREROS_k$ = relative erosion rate of pool k (1/yr)
∂t = time step (yr)

3.2.3 Mineral Weathering

Mineral weathering is considered for P and K. With respect to P, an inert, a stable and a labile inorganic solid phase pool are distinguished (Fig. III.1). Only the labile pool can supply P to solution according to first order kinetics (Flow 33). The primary source of P in the system is atmospheric deposition (Flow 28), which is directed through the stable pool (Flow 29), to resupply the labile pool (Flow 31). There is also a transfer flux from the labile to the stable inorganic pool of P (Flow 32). Part of the atmospheric deposition of P is stored in the inert pool (Flow 30), which will balance the P losses by erosion. Consequently, the total pool of inorganic inert P in the soil does not change in time. Per time step it holds:

$$RILSSP = INLAP \cdot (1 - \exp(-f_1 \cdot \partial t)) \qquad [III.26]$$

$$RILSP = INLAP \cdot (1 - \exp(-f_2 \cdot \partial t)) \qquad [III.27]$$

$$RISILP = INSTP \cdot (1 - \exp(-f_3 \cdot \partial t)) \qquad [III.28]$$

$$RDEP_P = REROSP_{ININP} + RDEPIS_P \qquad [III.29]$$

where:

RILSSP = rate of P transfer from soil inorganic labile P (INLAP) to solution (kg ha^{-1})
RILISP = rate of P transfer from INLAP to the soil inorganic stable P pool (INSTP) (kg ha^{-1})
RISILP = rate of transfer from INSTP to INLAP (kg ha^{-1})
$RDEPIS_P$ = replenishment of INSTP by atmospheric deposition (kg ha^{-1})
$RDEP_P$ = total atmospheric P deposition (kg ha^{-1})
$REROS_{ININP}$ – rate of erosion of the soil inorganic inert P pool (kg ha^{-1})
f_1, f_2, f_3 = relative rates of transfer for the respective transfers (RRILSS, RRILIS, RRISIL, respectively) (1/yr)
∂t = time step (yr)

For the time being, mineral weathering of K is modeled as a constant flux:

$$F_{wea} = ARWEA/n \qquad [III.30]$$

where:

F_{wea} = weathering flux per time step
ARWEA = annual rate of weathering (kg/ha per yr)
n = number of time steps per year

If simulations are carried out over thousands of years, or in case simulations include the addition of artificial small reactive pools of nutrients, depletion or increase of the mineral phase may be important. Examples of artificial pools are fertilizer addition and the presence of ash after burning. The effect of depletion could be accounted for by using a first order weathering model. Mineral weathering will increase too with decreasing pH. As pH is not simulated, such a dependency is not yet included. Changes in pH after fertilizer addition or burning may be appreciable. A general expression for the weathering flux (F_{wea}) is:

$$f_{wea} = k \cdot M/M_0 \cdot \exp(apH) \qquad [III.31]$$

where:

M_0	= initial mineral mass (kg ha^{-1})
M	= the actual mineral mass (kg ha^{-1})
a and k	= constants

A square root increase of silicate weathering rates with increasing hydrogen concentration is commonly reported (Stumm et al., 1985).

3.2.4 Cation exchange and adsorption

In the present model only adsorption of potassium is considered.
The exchange of potassium is described by linear adsorption (Fig. III.1: Flow 34):

$$ADS_K = K_d \cdot SSOL_k \qquad [III.32]$$

where:

ADS_K	= the adsorbed pool (kg/ha)
$SSOL_K$	= the aqueous pool (kg/ha)
K_d	= distribution coefficient

The adsorption equation is substituted in the mass balance equation (Section 3.5).

3.3 Organic pools and fluxes

3.3.1 Conversions and transfers

Organic matter is subject to two types of conversion processes: dissimilation into CO_2, H_2O and inorganic forms of P, N and K, and assimilation into microbial tissue. Conversion of organic P and N into inorganic forms is generally indicated as mineralization. Organic matter in the model is first subjected to conversion, whereafter the residues are transferred to a more stable organic pool. Eventually the organic residues will end up in the stable pool. The residence time of organic matter in all pools, except the stable organic pool, is one year. Conversion is brought about by the microbes, which utilize part of the converted elements for microbial growth. This process is also called immobilization. Depending on the quality of the organic substrate, conversion may lead to net immobilization of nutrients or net mineralization.

The conversion and transfer of leaf litter may require less than one year, e.g. in case of intensive biological activity or more than one year, e.g. under very acid or wet conditions (Staritsky, 1988). For wood litter much more time is required. At present, residence times are set at 1 and 15 yr, for leaf and wood litter, respectively. This implies for the model, that each year about 1/15 of the wood litter present is subjected to conversion and transfer. Data were derived from Noij (1988) and Vooren (1985). For fine-root debris and coarse-root debris, the same procedure is followed, with the same residence times of 1 and 15 yr, respectively.

After calculation of conversion and transfer fluxes for individual organic pools, fluxes are added to calculate the net element flux to solution and the net increase or decrease of the organic pools. The total dissimilation flux of carbon is not calculated because CO_2 leaves the soil system and is irrelevant in the simulation of forest growth.

The simulation of conversion of organic matter does not involve K. K is present in the primary organic pools as ion K^+. Most K will be released immediately after die-back of the vegetation compartment, the remainder is released fromthe primary organic pools (litter and root debris). Thus K transfer to the labile organic pool, which is the next stable pool in line, is negligible.

3.3.2 Calculation of dissimilation and transfer constants

3.3.2.1 Main principles

The calculation of the dissimilation constants is carried out only once. Dissimilation of organic matter is described according to Janssen (1984, 1986):

$$Y_t = Y_0 \cdot \exp\{4.7\,[(a + f \cdot \partial t)^{-0.6} - a^{-0.6}]\}$$ [III.33]

where:
Y_0 = the initial amount of organic matter (kg/ha)
Y_t = the remaining amount of organic matter (kg/ha)
∂t = time step (yr)
a = initial age of organic matter (yr)
f_t = correction factor for temperature

The value of f_t can be found by:

$$f_t = 2^{(T-9)/9}$$ [III.34]

where:
T = temperatur(°C)

Equation III.34 (modified from Janssen, 1986) predicts that the rate of dissimilation doubles for every 9°C increase of temperature. The age (a) is a measure of the stability of the organic pool. The increase in age of organic pools with time is equal to the product of f_t and the residence time of organic matter in the pool. For lowland tropical regions a value of 4 is taken for f_t. A value of 1 for f_t refers to organic matter dissimilation in temperate regions where the average annual temperature is 9°C. Equation III.34 predicts that decay in tropical regions where the average annual temperature amounts to 27°C, is four times faster than in temperate regions, as found by Jenkinson and Anayaba (1977). The residence time of organic matter in the labile and moderately labile pool is one year. Assuming an age of 2.18 years for the leaf litter pool (Janssen, 1984; Noij, 1988), the values of a for the labile and moderately labile pool would be 6.18 years and 10.18 years, respectively. However, the value of a for the organic stable pool is not 14.18 years, as the stable pool is an accumulation of organic matter of various ages. The a-parameter for the stable pool is generally calibrated by assuming that presently observed pools of stable organic matter are in steady state with present inputs or organic matter.

The procedure for transfer of organic matter is (see Fig. III.1):

- the leaf litter pool and fine root debris are transferred to the organic labile pool (Flow 12 and 17);
- a part of the wood litter and coarse root debris is transferred to the organic labile pool (Flow 13 and 15), the other part to the moderately labile pool (Flow 14 and 16);
- the residue of labile organic matter after one year is transferred to the organic moderately labile pool (Flow 18);
- the residue of moderately labile organic matter after one year is transferred to the organic stable pool (Flow 19).

3.3.2.2 Main pools and sub-pools

When using smaller time steps than a year, each organic pool (with exception of the stable pool) has to be further divided into a number of sub-pools equal to the number of time steps. The size of the first, least stable, sub-pool of the primary organic pools (Section 2.2) is the input from the vegetation die-back during one time step. The residue after dissimilation is transferred to the next sub-pool, etc. The residue of the last, most stable, sub-pool is transferred to the first sub-pool of the next in-line stable organic main pool.

Each sub-pool has its own "age" (a_i). At the beginning of the time step, the first sub-pool has the same age as the last sub-pool of the preceding pool at the end of the preceding time step (it is in fact the same material); this age is a. At the end of the time step the age of each sub-pool has increased by f/n years.
In formula:

$$a_{i,b} = a + (i - 1) \cdot f_t/n \qquad [III.35]$$

$$a_{i,e} = a + i \cdot f_t/n \qquad [III.36]$$

where:
$a_{i,b}$ = age of sub-pool i at the beginning of the time step
$a_{i,e}$ = age of sub-pool i at the end of the time step
a = age of the first sub-pool at the beginning of the time step (yr)
f_t = correction factor for temperature, see Equation III.34
n = the number of time steps per year

The masses of the sub-pools are calculated by Equation III.33, with the following values for the various parameters:
Y_t = Y_i, the mass of sub-pool i at the end of the time step (kg/ha)
Y_0 = mass of the first sub-pool at the beginning of the time step (kg/ha)
a = age of the first sub-pool at the beginning of the time step (yr)
∂t = (i-1)/n (yr)

The mass of the total pool is equal to the sum of the masses of the sub-pools. At the beginning of the time step, the mass of the first sub-pool is Y_0 and that of the last sub-pool is $Y_{(n-1)}$. At the end of the time step the masses are Y_1 and Y_n, respectively. The mass of the total pool decreases from $Y_{tot,b}$ to $Y_{tot,e}$ during the time step, where:

$$Y_{tot,b} = \sum_{i=1}^{n} Y_i \qquad [III.37]$$

$$Y_{tot,e} = \sum_{i=2}^{n+1} Y_i \qquad [III.38]$$

The rate of dissimilation of sub-pool i equals to the mass difference of the sub-pool between the beginning and the end of a time step:

$$J_{dis} = Y_i - Y_{i+1} \text{ (kg per time step)} \qquad [III.39]$$

The rate of transfer is equal to the mass of the sub-pool at the end of the time step:

$$J_{tra} = Y_{i+1} \text{ (kg per time step)} \qquad [III.40]$$

3.3.2.3 Dissimilation and transfer constants of main pools

Application of the procedure outlined above requires much calculation time. Therefore it was decided to apply the division into sub-pools to the leaf litter pool only, because the relative conversion rates of its sub-pools differ strongly. For the other main pools, conversions are described as first order reactions for the total pool. The sub-pool division is then used only to derive the first order rate constants. The first-order expression for dissimilation is:

$$dY/dt = - C_{dis} \cdot Y \qquad [III.41]$$

where C_{dis} = first order rate constant for dissimilation

Integration yields:

$$Y_t = Y_0 \exp(-C_{dis}t) \qquad [III.42]$$

$$Y_{tot,e} = Y_{tot,b} \exp(-C_{dis}t) \qquad [III.43]$$

So:

$$J_{dis} = Y_{tot,b} - Y_{tot,e} = Y_{tot,b} (1-\exp(-C_{dis}t)) \qquad [III.44]$$

Equation III.37 and III.38 give $Y_{tot,b}$ and $Y_{tot,e}$ from which, C_{dis} for the overall first-order dissimilation reaction can be calculated.

Transfer to a next main pool during one time step was calculated above as the mass of the last sub-pool at the end of a time-step (Eq. III.40). This transfer flux plus the dissimilation flux from the main pool (Eq. III.44) are used to derive a constant for the sum of the dissimilation and transfer fluxes (J_{dat}), which can be described similar to Eq. III.44:

$$J_{dat} = Y_{tot,b} (1 - \exp(- C_{dat}t)) \quad [III.45]$$

The transfer flux to the next main pool is then equal to:

$$J_{tra} = J_{dat} - J_{dis} \quad [III.46]$$

The initial ages of wood and coarse root debris that are transferred from the forest floor wood (FFW) pool and the coarse root debris (CRD) pool, respectively, is 8 years. The ages of the pools to which FFW and CRD are transferred are 6.18 (ORLA) and 10.18 (ORML) years, respectively (see Section 3.3.2.1). To overcome this age difference between transferred material and ORLA and ORML pools, a partitioning key, L, was introduced. This key brings about that the sum of dissimilation during one year of fraction L transferred to the ORLA-pool and of fraction (1-L) transferred to the ORML-pool equals the dissimilation during one year of eight-years old organic material (EYM).

$$J_{dis,ORLA} + (1-L) J_{dis,ORML} = J_{dis,EYM} \quad [III.47]$$

$J_{dis,ORLA}$, $J_{dis,ORML}$ and $J_{dis,EYM}$ can be calculated with Equation III.33 with t is 1 yr and ages of 6.18, 10.18 and 8 yr, respectively.

3.3.2.4 Dissimilation constant and age of the organic stable pool

For the calculation of the dissimilation constant of the organic stable pool a different procedure is used. The input in the stable pool per time step is the residue of the last sub-pool of the moderately labile pool: $J_{tra,OMRL}$ (EquationIII.46). As there is no transfer form the ORST pool, the following equation must hold in a steady state situation:

$$J_{tra,OMRL} = ORST \cdot (1 - \exp(- C_{dis.ORST}/n)) \quad [III.48]$$

where:
ORST = organic stable pool (kg/ha)
$C_{dis,ORST}$ = dissimilation constant of ORST (1/yr)
n = number of time steps

The C content of ORST is found as the total C content in the soil minus the C contents in the previously calculated pools. (If such calculations are not possible, it can be that assumed the C contents of the labile, moderately labile and stable pool are present in the proportion of 2:3:95). The unknown $C_{dis,ORST}$ in Equation III.48 can then be solved. At the end of the time step the size of the organic stable pool ($ORST_e$) is:

$$ORST_e = ORSTb \cdot \exp(- C_{dis,ORST}/n) \qquad [III.49]$$

The age of the organic stable pool (a) is found from:

$$ORST_e = ORST_b \cdot \exp(4.7 ((a + f/n)^{0.6} - a^{0.6}))) \qquad [III.50]$$

where a is the only unknown parameter.

3.3.3 Mineralization of N and P

The mineralization of N and P is calculated as the difference between conversion and assimilation of organic N and P. For that purpose the dissimilation-assimilation ratio (DA) during conversion must be known and the C-N and C-P ratios of the microbes involved (CN_m, CP_m). In the model these values are set at 2, 8.5 and 100 for DA, CN_m and CP_m, respectively (Noij, 1988). Such values refer to fungi grazing on organic matter with rather high C-N and C-P ratios like those present in tropical forests. The conversion of organic N and P follows from the conversion of carbon. The principles of the calculation are as follows.
The flux of converted carbon per time step (J_{convC}) is:

$$J_{conv,C} = J_{dis,C} + J_{ass,C} \qquad [III.51]$$

where $J_{dis,C}$ follows from Equation III.39 or Equation III.44, and

$$J_{ass,C} = J_{dis,C}/DA \qquad [III.52]$$

The flux of converted organic N and P per time step is:

$$J_{conv,N} = J_{conv,C}/CN_s \qquad [III.53]$$

$$J_{conv,P} = J_{conv,C}/CPs \qquad [III.54]$$

where CN_s and CP_s are the C-N and C-P ratios of the substrate that is converted. The assimilation of N and P is calculated by:

$$J_{ass,N} = J_{ass,C}/CN_m \qquad [III.55]$$

$$J_{ass,P} = J_{ass,C}/CP_m \qquad [III.56]$$

The mineralization fluxes of N and P then are;

$$J_{min,N} = J_{conv,N} - J_{ass,N}, \text{ and} \qquad [III.57]$$

$$J_{min,P} = J_{conv,P} - J_{assP} \qquad [III.58]$$

The values of $J_{min,N}$ and $J_{min,P}$ can be positive (net mineralization) or negative (net immobilization). The mineralization fluxes are indicated in Fig. III.1 by Flows 20-26.

At the end of each time step, the fluxes of dissimilated C, and mineralized N and P are subtracted from the quantities of these elements present at the beginning of each time step (C_0, N_0, P_0). This implies that the C-N and C-P ratios of the substrate change into:

$$CN_s = (C_0 - J_{disC})/(N_0 - J_{minN}), \text{ and} \qquad [III.59]$$

$$CP_s = (C_0 - J_{disC})/(P_0 - J_{minP}) \qquad [III.60]$$

The new values of CN_S and CP_S are used for the calculations in the following time step. The new ratios are lower than the original ones. In the course of the mineralization process the ratios further decrease, gradually approaching the C-N and C-P ratios of the microbes.

The total mineralization flux per year may depend on the time step. The smaller the time step, the more frequently the C-N and C-P ratios are adjusted (= lowered) and the higher the mineralization flux is. For time steps smaller than 0.005 yr, there is no further increase in mineralization flux. This time step is much smaller than that for the calculation of the other processes in the model, and its use would severely increase calculating time. Fortunately, there is a simple linear relation between the actual $J_{min,N}$ and the initial C-N, and between $J_{min,P}$ and the initial C-P ratio (Janssen and Noij, 1992; Noij, 1988).

$$J_{min,X} = (q/CR - r) \cdot C_0 \qquad [III.61]$$

where:

X	=	N or P
J_{min}	=	the actual mineralization flux during a time period
CR	=	the C-N or C-P ratio at the beginning of the time step
C_0	=	the carbon content of the pool at the beginning of the time step
q and r	=	regression coefficients

The values of q and r depend on the type of organic substrate, characterized by its "age" and on the length of the time step for which Equation III.61 is used.

In the unit DECCON of the model, q and r are determined for each organic pool, and for the desired time step of the model. For that purpose in DECCON two different CRs are used and a time step of 1/216. The resulting values of J_{minX} are substituted in Equation III.61 to solve q and r.

3.3.4 Release of K from organic pools

For potassium a different procedure is followed. In contrast to P and N, K is not structurally bound to C, but instead mainly present in the cell solution. Consequently, a large fraction (FFFLL,set at 0.5) of K will be released immediately after leaf litter fall (or fine root sloughing) and another fraction upon breakdown of the cell walls, proportionally to the C-dissimilation flux:

$$J_{rel,k} = J_{dis,c}/PORP_c \cdot PORP_k \qquad [III.62]$$

where:
$J_{rel,K}$ = K-release flux from primary organic pools (kg ha^{-1}) per time step
$J_{dis,C}$ = dissimilation flux of C (kg ha^{-1}) per time step
$PORP_C$ = amount of C in primary organic pool (kg/ha)
$PORP_K$ = amount of K in primary organic pool (kg/ha)

Potassium remaining in the last sub-pool, before transfer to the organic labile and moderately labile pool, is instantaneously released to solution at the end of the time step (Fig. III.1: Flow 11). The other K flows are not explicitly shown in Fig. III.1, to avoid too a complicated diagram.

3.4 Nutrient uptake and growth

3.4.1 General description and flow scheme

It should be emphasized that DYNAMITE does not explicitly simulate dry-matter production. Dry-matter production is related to uptake of N, P or K. It is assumed that in tropical regions solar radiation is not limiting forest growth.
The sequence of calculations during simulation of uptake and growth is (see also Fig. III.1) is not the same as the sequence of discussion in Sections 3.4.2 to 3.4.5. The calculations sub 2 to 5 are necessary because nutrient contents and concentrations and other vegetation characteristics have changed as a result of the fluxes in the preceding time step. The calculations sub 6 to 9 refer to the fluxes in the actual time step.

The sequence of calculations is:

1) in the first time step the element contents and concentrations for the various vegetation components and the total vegetation are initialized (cf. Section 4.2);
2) in the following time steps, the calculations start with the adjustment of substance contents and concentrations for total fine roots, coarse roots, wood and leaf (Section 3.4.5). The wood pool is further sub-divided into branches and stems. After adjustment, element concentrations are calculated for newly formed stems, branches, fine roots and coarse roots (Section 3.4.3);
3) calculation of the new leaf area index (Section 3.4.3.2);
4) calculation of the leaf and wood fall rate (Section 3.4.4);
5) calculation of fine root sloughing and transition of fine roots to coarse roots and calculation of coarse-root growth (Section 3.4.3.4);
6) calculation of the nutrient availability in solution. Determination of limiting nutrient and calculation of nutrient uptake per element. Reduction of nutrient uptake during soil moisture deficit (Section 3.4.2);
7) calculation of nutrient distribution between, and growth of, fine roots and leaves and calculation of Soil Fertility Index (SFI) (Section 3.4.3.2);
8) calculation of nutrients available for wood and calculation of wood growth (Section 3.4.3.3);
9) remaining nutrients, if any, are returned to soil solution. Calculation of net nutrient uptake by the vegetation, and recalculation of growth if necessary (Sections 3.4.3.3 and 3.4.4).

3.4.2 Soil nutrient availability and nutrient uptake

The maximum availability of nutrients from the soil solution is estimated from:

$$UP_{max,i} = SSOL_{i,b} + \Sigma\, in_i - NAN_i \qquad [III.63]$$

where:

$UP_{max,i}$ = maximum uptake of nutrient i (kg/ha)
$SSOL_{i,b}$ = amount of nutrient i in soil solution at the beginning of the time step (Section 3.5) (kg/ha)
$\Sigma\, in_i$ = sum of inputs of nutrient i from mineralization, weathering and atmospheric deposition during the time step
NAN_i = estimate of the amount of nutrient i in soil solution which is not available to plants during the time step (kg/ha).

NAN_i contains two terms. The first term is an estimate of the amount of nutrient i in soil solution which is not available due to leaching from the soil compartment. It is supposed that in case of drainage, only a fraction $(1-f_l)$ of $SSOL_{i,b}$ will be taken up by the vegetation during the time step, the other part (f_l) is supposed to leach during the time step. The second term is an estimate of nutrient i in soil solution which is not available, because its concentrations is below the minimum concentration for root uptake.

$$NAN_i = f_l \cdot J_{out}/\Theta \cdot D \cdot SSOL_{i,b} + 10\ \Theta \cdot D \cdot C_{min,i} \qquad [III.64]$$

where:

$C_{min,i}$ = minimum concentration of nutrient i in soil solution required for root uptake (kg/m^3)

f_l = leaching fraction

When there is a soil moisture deficit, nutrient uptake will be reduced according to the procedure used for reducing root water uptake (Equation III.15):

$$Up_{act,i} = UP_{max,i} \cdot (\Theta - \Theta_{wp})/(\Theta_{ru} - \Theta_{wp}) \qquad [III.65]$$

$$\Theta_{ru} = \Theta_r \cdot f_{ru} \qquad [III.66]$$

where:

$UP_{act,i}$ = actual uptake of nutrient i (kg/ha)

$UP_{max,i}$ = uptake of nutrient i based on nutrient availability

Θ_{ru} = water content below which reduction in nutrient uptake starts

Θ_r = water content below which reduction in water uptake occurs

f_{ru} = constant, presently set to 1

As explained in Part II, Section 4, only the most limiting nutrient is taken up as calculated with $UP_{act,i}$. The uptake of the other nutrients depend on their ratio to the most limiting nutrient.

The ratios N/P, P/K and K/N for uptake are calculated from the corresponding ratios in the soil solution, according to empirical power relationships (derived in Part II, Section 4).

$$\log (RATX, Y_{up}) = a + b \cdot \log (RATX, Y_{ss}) \qquad [III.67]$$

where:

$RATX,Y_{up}$ = ratio of nutrient X to nutrient Y in plant uptake flux

$RATX,Y_{ss}$ = ratio of available amount of nutrient X to that of nutrient Y in soil solution during the time step; available amount as caculated with $UP_{max,i}$

a,b = intercept and slope

Furthermore, uptake ratios should satisfy:

$$RATX,Y_{up,min} < RATX,Y_{up} < RATX,Y_{up,max} \qquad [III.68]$$

$RATX,Y_{up,min}$ and $RATX,Y_{up,max}$ refer to the minimum and maximum values the ratio can have. The concepts are described in Part II, Section 4. Next the ratios $RATX,Y_{up}/RATX,Y_{ss}$ are calculated for all possible combinations of XY. The nutrient X giving the highest value of that ratio is the most limiting nutrient. The uptake for the other nutrients follows from:

$$UP_{act,Y} = UP_{act,X_l} \cdot RATY,X_l \qquad [III.69]$$

with X_l = limiting nutrient for total uptake.

3.4.3 Nutrient distribution in and growth of the vegetation

3.4.3.1 General principles

The growth of a plant component is calculated by:

GROWTH = NUTRIENT UPTAKE/NUTRIENT CONCENTRATION

This means that for calculation of growth the uptake and concentration of nutrients in the newly grown part of a plant component must be known.

The total uptake of nutrients is calculated in Section 3.4.2. Initially, the nutrients that are taken up are distributed between fine roots and leaves (Fig. III.1: Flow 1, 2 and 3). The distribution is a function of the nutrient concentrations in fine roots and the maximum and minimum concentrations in leaves. The remainder of the nutrients, if any, are sent to wood. The concentration of nutrients in newly grown roots and wood are functions of the nutrient concentration in leaves at the beginning of the time step.

The actual growth of a plant component is determined by the limiting nutrient. The required uptake of the other nutrients then follows from:

UPTAKE = GROWTH · CONCENTRATION

3.4.3.2 Fine roots and leaves

The total uptake of nutrient i ($UP_{act,i}$) is distributed between the newly formed fine roots and the newly formed leaves:

$$UP_{act,i} = GROWFR \cdot FRNC_i + GROWL \cdot LNC_i \qquad [III.70]$$

where:

$GROWFR$ = dry matter in newly formed fine roots (kg/ha)
$GROWL$ = dry matter in newly formed leaves (kg/ha)
$FRNC_i$ = concentration of nutrient i in the newly formed fine roots (kg/kg)

LNC_i = concentration of nutrient i in newly formed leaves (kg/kg)

The concentration of nutrient i in the newly formed fine roots is calculated as a function of the concentration in leaves at the beginning of the time step ($LEAFC_{i,b}$).

$$FRNC_i = a_i + b_i \cdot LEAFC_{i,b} \qquad [III.71]$$

The values of a_i and b_i follow from empirical relationships (Part II, Section 5).

Next, it is determined which nutrient would be limiting for fine root growth, if the whole quantity of nutrients taken up would be allocated to fine roots. The limiting nutrient (il) is the one for which the ratio $UP_{act,i}/FRNC_i$ is the lowest. It is not necessary the same nutrient that was limiting for the total uptake. It is assumed that the concentration of this nutrient in the newly formed leaves will have the minimum value (Part II, Table 8), indicated by $LEAFCD_{il}$. (D stands for dilution). Substituting this value in Equation III.70 and reorganizing gives the following relationship between GROWFR and GROWL:

$$GROWFR = UP_{act,il}/FRNC_{il} - GROWL \cdot LEAFCD_{il}/FRNC_{il} \qquad [III.72]$$

$$GROWFR = A - B \cdot GROWL \qquad [III.73]$$

Substitution of Equation III.73 in Equation III.70 yields:

$$UPT_{act,i} = (A - B \cdot GROWL) \cdot FRNC_i - GROWL \cdot LNC_i \qquad [III.74]$$

For each of the non-limiting nutrients, a range of possible leaf growth can be calculated. These ranges are between $GROWLA_i$ and $GROWLD_i$.

$$GROWLA_i = (UP_{act,i} - A - FRNC_i)/(LEAFCA_i - B \cdot FRNC_i) \qquad [III.75]$$
$$GROWLD_i = (UP_{act,i} - A \cdot FRNC_i)/LEAFCD_i - B \cdot FRNC_i) \qquad [III.76]$$

where:

$LEAFCA_i$ = maximum (A stands for accumulation) concentration of nutrient i in leaves

$LEAFCD_i$ = minimum (D stands for dilution) concentration of nutrient i in leaves

Usually the ranges found for the various nutrients, have a common overlap between $GRLOWLA_{max}$ and $GRLOWLD_{min}$, where:

$GROWLA_{max}$ = the maximum $GROWLA_i$
$GROWLD_{min}$ = the minimum $GROWLD_i$

The actual leaf growth (GROWL) is set equal to the mid point of the common overlap:

$$GROWL = 0.5\,(GROWLD_{min} + GROWLA_{max}) \qquad [III.77]$$

If there is no overlap, in other words if $GROWLD_{min} < GROWLA_{max}$, the actual leaf growth is:

$$GROWL = GROWLD_{min} \qquad [III.78]$$

GROWL should be lower than the maximum leaf growth ($GROWL_{max}$) as determined by the transpiration (T_{act}; Section 3.1.7) and transpiration ratio (TRR):

$$GROWL_{max} = T_{act}/TRR \qquad [III.79]$$

where:
TRR = the minimum amount of water required per kg of leaf dry-matter production (kg/kg).

The ratio $GROWL/GROWL_{max}$ is called Soil Fertility Index (SFI). The value of SFI is between 0 and 1. If GROWL is more than $GROWL_{max}$, water availability limits growth stronger than nutrient availability.

The new growth of fine roots is found by substitution of GROWL, calculated with Equation III.77, III.78 or III.79, in Equation III.69. Subsequently, the uptake of nutrients in fine roots ($UPFR_i$) is calculated by

$$UPFR_i = GROWFR \cdot FRNC_i \qquad [III.80]$$

and the uptake of nutrients (UPL_i) in leaves by

$$UPL_i = UP_{act,i} - UPFR_i \qquad [III.81]$$

Nutrients can be stored in the leaf component up to a maximum concentration:

$UPLA_i$ = $GROWL/LEAFCA_i$ [III.82]

$UPL_{act,i}$ = MINIMUM (UPL_i, $UPLA_i$) [III.83]

where:

$UPLA_i$ = uptake of i at maximum concentration of i in leaves

$UPL_{act,i}$ = actual uptake of i in leaf

$LEAFCA_i$ = maximum leaf element concentration of nutrient i

MINIMUM (X, Y) = function selecting smallest value of X and Y

If $UPL_i > UPL_{act,i}$, the difference ($UPL_i - UPL_{act,i}$) is sent to the wood (Fig. III1: Flow 5). It is considered as the direct uptake of nutrient i by wood (Section 3.4.3.3). Besides, there is element translocation from leaf to wood prior to litter fall. Its rate ($RLWRE_i$; Fig. III.1: Flow 6) is a fraction ($FLRDU_i$) of the total amount of i ($LEAF_i$) in leaf subject to fall.

$$RLWRE_i = LEAF_i \cdot FLFA \cdot FLRDU_i \qquad [III.84]$$

where:

FLFA = fraction of leaf mass which will fall (See Section 3.4.4)

$FLRDU_i$ = reduction fraction of i in leaves before leaf fall.

The values of $FLRDU_i$ for K and dry matter are fixed, those for N and P are a function of (i) the regular leaf element concentration ($LEAFC_i$), (ii) the lowest possible leaf element concentration just before fall ($LEAFLC_i$):

Reduction fractions for individual elements are different. They are calculated by:

$$FLRDU_i = MINIMUM(Y_1, Y_2) \qquad [III.85]$$

where:

$$Y_1 = (LEAFC_i - LEAFLC_i)/LEAFC_i) \qquad [III.86]$$

$$Y_2 = c + d \cdot LEAFC_i \qquad [III.87]$$

where c and d are fixed coefficients. For explanation and values of coefficients, see Part II, Section 6.

For dry matter $RLWRE_i$ is not calculated. The calculation of $FLRDU_{DM}$ is needed for the assessment of the amount of dry matter in falling leaves.

Transition of fine roots to coarse roots is discussed in Section 3.4.3.4.

3.4.3.3 Wood

The remaining nutrients after uptake and growth of the fine roots and leaves are available for the wood (UPW_i; Fig. III.1: Flow 5):

$$UPW_i = UP_{act,i} - UPFR_i - UPL_{act,i} \qquad [III.88]$$

This amount of nutrients is supplemented with nutrients from redistribution in the leaf compartments before leaf litter fall ($RLWRE_i$, Equation III.84). The total rate of nutrients entering wood ($RENW_i$) then becomes:

$$RENW_i = UPW_i + RLWRE_i \qquad [III.89]$$

Similar to calculations for the other vegetation components, wood growth is calculated as:

$$GROWW = MINIMUM\ (RENW_i/WOODNC_i) \qquad [III.90]$$

where:

$MINIMUM\ (X_i)$ = minimum value of X_i for i = N, P or K

$WOODNC_i$ = concentration (mass fraction) of nutrient i in newly formed wood

The remaining nutrients ($SUPW_i$), if any, are returned to solution.

$$SUPW_i = RENW_i - (GROWW \cdot WOODNC_i) \qquad [III.91]$$

When calculating the nutrient concentration in newly formed wood, distinction is made between branch wood and stem wood. Nutrient concentration in stem wood and branch wood are both linearly depending on the leaf element concentrations.

$$BRNC_i = a + b\ LEAFC_{i,b} \qquad [III.92]$$

$$STEMNC_i = c + d\ LEAFC_{i,b} \qquad [III.93]$$

where:

$BRNC_i \leq BRANCA_i$

$STEMNC_i \leq STEMCA_i$

$BRANCA_i$ = maximum concentrations of nutrient i in branches (Part II, Section 5).

$STEMCA_i$ = maximum concentration of nutrient i in stems

A fixed fraction of wood growth is allocated to branches (FBRA) and stem (1-FBRA). The average concentration of nutrient i in the wood compartment is thus calculated as:

$$WOODC_i = FBRA \cdot BRANC_i + (1 - FBRA) \cdot STEMC_i \qquad [III.94]$$

3.4.3.4 Coarse roots

Turnover of fine roots (RFRTR), in kg/ha per time step comprises two flows:
- transition of fine roots to coarse roots (RFRCR); Fig. III.1: Flow 4), and
- sloughing of fine roots to fine root debris (RFRSL; Fig. III.1: flow 10).

Total turnover is described by:

$$RFRTR = FROOT_{DM} \cdot (1 - \exp (RRFRTR \cdot \partial t) \qquad [III.95]$$

where:
$FROOT_{DM}$ = dry mass of fine roots (at the beginning of the time step) (kg/ha)
RRFRTR = relative rate of fine-root turnover (1/yr)

The rate of transition from fine to coarse roots (RFRCR, in kg/ha per time step) is equal to:

$$RFRCR = FFRCRT \cdot RFRTR \qquad [III.96]$$

where:
FFRCRT = fraction of rate of fine-root turnover that is translocated to coarse roots

In Part II, Section 2.4, it is explained that FFRCRT is calculated by:

$$FFRCRT = (1 - \exp (- RRCRTR))/(1 - \exp (-RRFRTR)) \cdot q_r \cdot SFI \cdot MIN(CRNCi/FRNC_i) \qquad [III.97]$$

where:
RRCRTR = relative rate of coarse-root turnover (1/yr), which is related to the relative rate of woodfall, RRWFA via Eq. III.114
q_r = ratio of $CROOT_{DM}/FROOT_{DM}$ to SFI (see Part II, Section 2.4)
SFI = Soil Fertility Index, calculated in Section 3.4.3.2
CRNCi = concentration of nutrient i in newly grown coarse roots (kg/kg)

$$SFI = GROWL/GROWL_{max} \qquad [III.98]$$

$CRNC_i$ and $FRNC_i$ are calculated by Equations III.96 and III.67, respectively. The transfer fluxes for individual elements ($RFRCR_i$) are obtained by multiplying RFRCR by the respective element concentrations ($FROOTC_i$).

$$RFRCR_i = RFRCR \cdot FROOTC_i \qquad [III.99]$$

where:

$FROOTC_i$ = the concentration of element i in total fine roots.

The nutrient concentration in the newly formed coarse roots ($CRNC_i$) is calculated as the average of the nutrient concentrations in newly formed stems and newly formed fine roots:

$$CRNC_i = 0.5\,(STEMNC_i + FRNC_i) \qquad [III.100]$$

The growth of the coarse roots is calculated in the same way as shown for wood.

$$GROWCR = MINIMUM\,[(RFRCR_i)/CRNC_i] \qquad [III.101]$$

where i stands for N, P or K.

Remaining nutrients ($SUPCR_i$), if any, are returned to solution:

$$SUPCR_i = RFRCR_i - GROWCR/CRNC_i \qquad [III.102]$$

3.4.4 Leaf and wood fall, fine-root sloughing and coarse-root turnover

Litter fall and root turnover are calculated as fractions of the dry matter present in the relevant vegetation parts at the beginning of the time step.

The leaf fall rate (RLFA) is:

$$RLFA = FLFA \cdot LEAF \qquad [III.103]$$

$$FLFA = 1 - \exp(-RRLFA \cdot \partial t) \qquad [III.104]$$

where:

FLFA = fraction of leaves which will fall per time step; FLFA is smaller than or equal to 1

RRLFA = relative rate of leaf fall (1/yr)

The relative rate of leaf fall is calculated as a function of the concentration of nutrients, but this function might require correction for light stress (f_{LAI}) or moisture stress (f_{MS}) (Part II, Section 7.3). The relative rate of leaf fall, as determined by nutrient i only, is calculated as follows.

$$RRLF_i = r + s \cdot (LEAFC_i - A_i/B_i) \qquad [III.105]$$

where:

$RRLF_i$ = relative rate of leaf fall, as determined by nutrient i (1/yr); its value should lie between 0.8 and 2.0
r,s = regression constants
A_i, B_i = constants, values of $LEAFC_i$ (kg/kg)

The multiplication factor for light stress (caused by a high LAI) is calculated as follows.

$$f_{LAI} - q_l \cdot (LAI - LAI_{cr})/LAI_{cr} \qquad [III.106]$$

where:

q_l = constant, presently set at 3
LAI_{cr} = critical LAI; for tropical forests its value is assumed to be 4

The multiplication factor for moisture stress (f_{MS}) is calculated as follows:

$$f_{MS} = q_m (T_{pot} - T_{act})/T_{pot} \qquad [III.107]$$

where:

q_m = constant, presently set at 3
T_{pot}, T_{act} = potential and actual transpiration

The values of f_{LAI} and of f_{MS} should lie between 0 and 3.
Finally, the relative rate of leaf fall (RRLFA) is calculated as the product of the lowest value of $RRLF_i$ (for i = N, P or K), and the highest value of f_{LAI} and f_{MS}, provided their values are more than 1.

$$RRLFA = MINIMUM (RRLF_i) \cdot MAXIMUM (1, f_{LAI}, f_{MS}) \qquad [III.108]$$

To calculate the rate of transfer of nutrients by leaf fall, the nutrients that are redistributed have to be substracted from the product of leaf fall and leaf nutrient concentration:

$$RLFA_i = LEAF_i \cdot RLFAR - RLWRE_i \qquad [III.109]$$

where:

$LEAF_i$ = total amount of nutrient i in leaves (kg/ha)

$RLWRE_i$ = rate of nutrient redistribution (kg/ha), as calculated in Equation III.80.

For $RLFA_{DM}$ (rate of transfer of dry matter by leaf fall) it holds:

$$RLFA_{DM} = (1 - FLRDU_{DM} \cdot LEAF_{DM} \cdot RLFAR \qquad [III.110]$$

where:

$FLRDU_{DM}$ = reduction fraction for DM in leaves before leaf fall.

The relative rate of wood fall (RRWFA, 1/yr) is a discontinuous function of the wood mass. The value of the function is determined by linear interpolation in a function AFGEN ($WOOD_{DM}$, RRWFA), modified after Noij (1988). The rate of wood fall during a time step is calculated as follows.

$$RWFA = FWFA \cdot WOOD_{DM} \qquad [III.111]$$

where:

RWFA = rate of wood fall (kg/ha per time step)

FWFA = fraction of wood fall per time step

$WOOD_{DM}$ = wood dry matter (kg/ha)

$$FWFA = 1 - \exp (RRWFA \cdot t) \qquad [III.112]$$

$$RRWFA = AFGEN (WOOD_{DM}, RRWFA) \qquad [III.113]$$

The rate of nutrients transferred by wood fall ($RFWA_i$) (Fig. III.1: Flow 8) is:

$$RFWA_i = RFWA \cdot WOOD_i \qquad [III.114]$$

where $WOOD_i$ = total amount of nutrient i in wood (kg/ha)

The rate of fine root sloughing (RFRSL) is (Part II, Section 2):

$$RFRSL = (1\text{-}FFRCRT) \cdot RFRTR \qquad [III.115]$$

RFRTR and FFRCRT are calculated by the Equations III.95 and III.97, respectively.

The rate of nutrient i transferred by fine-root sloughing to fine-root debris ($RFRSL_i$) (Fig. III.1: Flow 10) is:

$$RFRSL_i = RFRSL \cdot FROOT_i \qquad [III.116]$$

where $FROOT_i$ = total amount of nutrient i in fine roots (kg/ha)

The rate of coarse-root turnover (RCRTR) is a fixed fraction (FCRTR) of the coarse-root mass:

$$RCRTR = FCRTR \cdot CROOT_{DM} \qquad [III.117]$$

where:
$CROOT_{DM}$ = dry mass of coarse roots (kg/ha)

$$FCRTR = f_{CRTR} \cdot WFA \qquad [III.118]$$

where:
f_{CRTR} = constant, presently set at 2

The rate of nutrient i transferred by coarse-root turnover to coarse-root debris, $RCRTR_i$ (Fig. III.1: Flow 9), is:

$$RCRTR_i = RCRTR \cdot CROOT_i \qquad [III.119]$$

where $CROOT_i$ = total amount of nutrient i in coarse roots.

3.4.5 Net growth and adjustment of vegetation components

The net increase of nutrient contents for the various vegetation components is:

$$NGROWFR_i = GROWFR_i - RFRSL_i - RFRCR_i \qquad [III.120]$$

$$NGROWCR_i = GROWCR_i - RCRTR_i \qquad [III.121]$$

$$NGROWL_i = GROWL_i - RLFA_i - RLWRE_i \qquad [III.122]$$

$$NGROWW_i = GROWW_i - RWFA_i \qquad [III.123]$$

where:
prefix N = net
i = N, P, K, or DM

For the increase in leaf dry matter the calculation is a little different than in Equation III.122, namely

$$NGROWL_{DM} = GROWL_{DM} - LEAF_{DM} \cdot RLFAR \qquad [III.125]$$

This implies that total DM, originally present in the leaves that will fall, is subtracted from the gross increase in leaf DM.

Next the final amounts of nutrients and DM in the various vegetation components are calculated, i.e. the amounts present at the beginning of the following time step, e.g.

$$FROOT_i = FROOT_i + NGROWFR_i \qquad [III.125]$$

The nutrient concentrations are found by e.g.:

$$LEAFC_i = LEAF_i / LEAF_{DM} \qquad [III.126]$$

This $LEAFC_i$ is the concentration at the beginning of the next time step ($LEAFC_{i,b}$), to which are related the nutrient concentrations in fine roots, branches and stems that will be formed during the next time step (Eq. III.71, III.92 and III.93)

Further the new leaf area index is calculated according to:

$$LAI = SLA \cdot LEAF_{DM}/10000 \qquad [III.127]$$

SLA is found as the average value of SLA calculated as a function of leaf N, and SLA calculated as a function of leaf P:

$$SLA = 0.5 \cdot (q + r \cdot LEAFC_N) + s + t \cdot LEAFC_P) \qquad [III.128]$$

with

SLA	= specific leaf area ($m^2\ kg^{-1}$)
$LEAF_{DM}$	= leaf dry matter ($kg\ ha^{-1}$)
$LEAFC_N$	= N concentration in leaf (g/g)
$LEAFC_P$	= P concentration in leaf (g/g)
q, r, s, t	= constants, presently set at 1.99, 405, 2.51 and 6804, respectively

3.5 Chemical balance and adjustment of pools

For the calculation of the total amount of nutrient i is the soil solution and in the adsorbed pool at the end of the time step, which comes down to the adjustment of these pools at the beginning of the next time step, the following formulation for the chemical balance is used:

$$SSOL_{i,e} + ADS_{i,e} = SSOL_{i,b} + ADS_{i,b} + \Sigma_{in,i} - UP_{act,i} - F_{out,i} \qquad [III.129]$$

with:

$SSOL_{i,b}$ = amount of nutrient i in soil solution at the beginning of the time step
$SSOL_{i,e}$ = amount of nutrient i in soil solution at the end of the time step
$ADS_{i,b}$ = amount of adsorbed nutrient i at the beginning of the time step
$ADS_{i,e}$ = amount of adsorbed nutrient i at the end of the time step
$\Sigma\ in_i$ = sum of net inputs into soil solution of nutrient i during the time step
$UP_{act,i}$ = total actual uptake of nutrient i during the time step
$F_{out,i}$ = leaching of nutrient i with the outgoing waterflux during the time step

The individual terms are calculated as follows.

$$ADS_i = Kd \cdot SSOL_i \qquad [III.130]$$

where:

K_d linear distribution coefficient for element contents in the adsorbed and aqueous phase (Section 3.2.4)

Σin_i consists of the following fluxes.

$$\Sigma in_i = F_{dep,i} + F_{wea,i} + F_{min,i} + F_{in,i} \qquad [III.131]$$

$F_{dep,i}$ = atmospheric deposition of nutrient i
$F_{wea,i}$ = mineral weathering of nutrient i
$F_{min,i}$ = mineralization of nutrient i from all organic matter pools
$F_{in,i}$ = input of nutrient i from overlying soil layer

F_{out} is calculated with the average element concentration and water content at the beginning and the end of the time step.

$$F_{out} = J_{out} \cdot SSOL_{i,b} + SSOL_{i,e}/D \cdot (\Theta_b + \Theta_e) \qquad [III.132]$$

This gives the following explicit expression for $SSOL_{i,e}$:

$$SSOL_{i,e} = SSOL_{i,b} \cdot (K_d + 1 - a) + \Sigma\ in_i - U\ P_{act,i}/K_d + a + 1 \qquad [III.133]$$

where:

$$a = J_{out}/D \cdot (\Theta_b + \Theta_e) \qquad [III.134]$$

The other pools are adjusted in a similar way, but the formulation is simpler because no water flux is involved:

$$POOL_{k,i} = POOL_{k,i} + \Sigma\, in_{k,i} - \Sigma\, out_{k,i} \qquad [III.135]$$

where:

$POOL_{k,i}$ = amount of substance i in pool k (kg/ha)

$\Sigma\, in_{k,i}$ = sum of inputs of nutrient i into pool k during time step (kg/ha)

$\Sigma\, out_{k,i}$ = sum of outputs of nutrient i from pool k during time step (kg/ha)

4 MODEL INPUT

4.1 Introduction

A general description of required input data for the DYNAMITE simulation model is given in Section 3.3, Part I. Table III.2 shows the required hydrologic input data and Table III.3 the chemical and biological input parameters. The values of the data apply to a tropical forest in the Taï region, Côte d'Ivoire, unless stated otherwise.

With respect to the required input data, three parameter types have been distinguished:

- initial values for state variables; divided into state variables for the soil components and state variables for the vegetation components;
- intrinsic system variables, extracted from the intrinsic relations described in Part II; subdivided following the sequence of the discussion in Section 3;
- boundary fluxes or boundary flux regulating variables (e.g. deposition fluxes and relative erosion rate).

4.2 Input file for the moisture cycling sub-model

See Table III.2

- Average monthly rainfall (AMR) and pan evaporation (AME_{pan}) data for the Taï area were taken form Collinet et al., 1984. Days with rain within a month (DAYSWR) are estimated values.

- All nutrient related processes take place in the first soil layer. Its thickness is set at 200 mm. Water uptake can also occur form the second soil layer. At present we suppose that 40% of the potential water uptake can take place from the second soil layer (WUF in Eq. III.13). The litter layer is not considered in the water balance calculations.

The volume fractions of moisture (Θ) and gas are all estimated values. The symbols refer to the moisture fraction at the beginning of the simulations period (Θ), at saturation (Θ_s), at field capacity (Θ_{fc}), at wilting point (Θ_{wp}), and after air drying (Θ_{ad}), and to the minimum volume fraction of soil gas (d). Measured soil physical data are not yet available for the Taï-area. The maximum capillary rise (J_{cr}, mm/day) was obtained from Table III.1, loamy sand.

Table III.2 The input file (WATER.INP) for the water balance sub-model of DYNAMITE

Month	AMR mm/month	AME_{pan} mm/month	DAYSWR
January	21	110	4
February	65	114	4
March	148	130	8
April	170	124	10
May	215	116	16
June	269	80	22
July	124	82	14
August	132	80	8
September	293	82	12
October	240	110	14
November	108	116	14
December	47	110	8

Number of soil layers: 2
Thickness of soil layers (mm): 200, 800
Relative distribution of water uptake over soil layer 1 and 2: 0.6, 0.4

Layer	Θ	Θ_s	Θ_{fc}	Θ_{wp}	Θ_{ad}	d	J_{cr}	a
1	0.24	0.40	0.25	0.05	0.04	0.05	1.84	44.5
2	0.24	0.40	0.25	0.05	0.04	0.05	2.02	51.7

Crop factor, f_1: 1.15
Correction of potential evapotranspiration for interception, f_2: 0.9
Parameters used in the interception model:
a: $5.3214.\ 10^{-5}$ b: 0.17959

- The crop factor f_1 is used to calculate Penman evapotranspiration from pan evapotranspiration (Eq. III.8). The daily pan evaporation in the Taï region ranges from 2.5-4.0 mm/day. For such a range, Poels (1987) gives an average value of 1.15 for f_1.

- f_2 is a factor in the equation (Eq. III.9) that corrects the Penman potential evapotranspiration for direct evaporation of intercepted canopy water. The value of 0.9 was taken from Poels (1987).

- The parameters a and b are used in the interception calculations (Eq. III.4). These values should be calibrated beforehand.

4.3 Input file for the nutrient cycling sub-model

Initial values for state variables: the soil components

See Table III.3.

- In Part I, Section 3.3.3.2, it is discussed how initial values for inorganic pool contents for phosphorus (ININP,INSTP,INLAP) can be derived from chemical soil data.

- Organic pool contents for nitrogen, phosphorus, potassium and carbon. If a model run starts from a steady-state situation the contents of the organic labile (ORLA) and organic moderately labile (ORML) pools follow from the transfer rates of the primary organic pools (FFW, FFL, FRD, CRD) in the initialization procedure in the program (procedure DECCON, see Appendix III.2). In this case, the ORLA and ORML contents as read from Table III.3 are ignored and the content of the organic stabile pool (ORST), as read from the input file is interpreted as total content. In the initialization procedure the actual content of ORST is then calculated from total soil content minus the contents of the other organic pools in the mineral layer.
  If the steady-state situation is not known, the ORLA, ORML and ORST contents are derived directly from the contents as read from the input file. So, in this case the ORST content, as read from the input file, is not the total soil content but the actual ORST content.

- Total N and P content of the soil were estimated from data by Fritsch (1982).

- In the model runs until now, initial solution contents ($SSOL_N$, $SSOL_P$, $SSOL_K$) were estimated.

Table III.3 Initial values of state variables (kg/ha)

State variable		Elements or dry (organic) matter (DM)				
Description	Symbol	N	P	K	C	DM
Soil						
inorganic inert	ININ		41.65			
inorganic stable	INST		70.85			
inorganic labile	INLA		7.77			
organic stable	ORST	1898.5	158.66		27000	
org. mod. labile	ORML	66.3	5.40		1270	
organic labile	ORLA	46.5	3.73		863	
soil solution	SSOL	4.7	0.2	4.2		
Forest floor (litter layer) and root debris						
leaf	FFL	3.8	0.72	8.1	787	
wood	FFW	106.0	20.14	218.0	36940	
fine root	FRD	9.8	0.29	5.9	1343	
coarse root	CRD	4.7	4.55	42.8	7462	
Vegetation						
leaf	LEAF	16.5	2.88	31.2		2650
wood	WOOD	711.0	135.00	1462.0		450000
fine root	FROOT	17.2	1.65	15.0		3000
coarse root	CROOT	320.0	30.50	287.0		50000

Initial values for state variables: the vegetation components

- The data refer to a situation in which the forest ecosystem is in steady-state. The dry matter in the above- ground vegetation is set at 450 tons/ha (Huttel, 1977). The amounts and concentrations of nutrients in the vegetation components were derived from model runs over a period of 3000 years, which gave a wood dry matter of content of about 450 tons/ha at steady-state.
  After initializing the model from steady state, growth following clearcutting can be simulated. Then contents of vegetation components are set to values as shown in Table I.1 in Section 3.4.2 of Part I.

Intrinsic system variables

See Table III.4.

- Inorganic phosphorus transfer processes were calculated according to Wolf et al. (1987), as discussed in Part I, Section 2.2.2.1.

Table III.4 Values of intrinsic system parameters

Relative rates of transfer of inorganic P (1/yr)

RRISIL: 0.034 RRILIS: 0.223 RRILSS: 0.105

Distribution coefficient of adsorbed K and K in solution (K_d): 100

Relative rates of removal of primary organic pools (1/yr)

RRFFLR	1.0	RRFRDR	1.0
RRFFWR	0.0667	RRFRCR	0.0667

Initial age (a) of organic pools (yr)

Forest floor leaves (FFL)	2.18	Soil organic pools	
Forest floor wood (FFW)	4.00	Org. labile (ORLA)	6.18
Fine-root debris (FRD)	2.18	Org. mod. lab. (ORML)	10.18
Coarse-root debris (CRD)	4.00	Org. stable (ORST)	25.19

Properties of micro-organisms

DA: 2 CN_m: 8.5 CP_m: 100 NP_m: 11.8

Temperature correction factor (f_r): 4 (for temp. 27° C)

Ratio of carbon to dry matter (CDM): 0.5

Fraction of K immediately leached from fine litter (kg/kg)

FFFLL: 0.5 FFRDL: 0.5

Fraction of $SSOL_{i,b}$ that is leached during time step (f_l): 0.5

Minimum concentration for uptake from solution (C_{min}) (kg/m^3)

N 1.0 10^{-3} P 6.0 10^{-6}

Logarithmic regression equation for nutrient uptake ratios

	UN/UP	UP/UK
Intercept	8.3	-0.05
Slope	0.7	0.95

Minimum and maximum values of uptake ratios

	N/P	P/K	K/N
Minimum	3.0	0.04	0.26
Maximum	20.0	0.6	2.6

Ratio of water volume fraction below which reduction in nutrient uptake starts to water volume fraction below which reduction in water uptake starts (f_{ru})

N 1.0 P 1.0 K 1.0

Minimum and maximum concentrations of nutrients in leaves

	N	P	K
Minimum	0.0075	0.0003	0.0040
Maximum	0.0250	0.0020	0.0200

Fraction of wood allocated to branches (FBRA): 0.33

Parameters for the equations relating nutrient concentrations in fine roots, branches or stems to nutrient concentrations in leaves (kg/kg). Min = minimum concentrations, Max = maximum concentrations.

		Intercept	Slope	Min.	Max.
Fine roots	N	0.00186	0.441	0.005	0.014
	P	-0.0007	0.622	0.00012	0.0013
	K	-0.00539	1.255	0.001	0.0075
Branches	N	-0.001	0.31	0.002	0.007
	P	-0.00085	1.0	0.00015	0.00115
	K	0.00106	0.213	0.0019	0.005
Stems	N	0.00034	0.115	0.0012	0.0032
	P	-0.0004	0.5	0.0001	0.0006
	K	0.00007	0.243	0.001	0.0050

Transpiration ratio (TRR): 300 kg water/kg dry matter

Parameters for the calculation of nutrient reduction fraction ($FLRDU_i$)

	lowest concentration in fallen leaves (kg/kg)	intercept	slope
N	0.0051	0.85	-20.0
P	0.00015	0.78	-166.7
K	fixed value for $FLRDU_i$:	0.15	
DM	fixed value for $FLRDU_i$:	0.28	

Relative rate of fine-root turnover RRFRTR (1/yr): 1.0

Parameters for the calculation of the relative rate of leaf fall (RRLFA)

- calculation of $RRLF_i$

	r	s	A	B
N	0.8	1.2	0.010	0.010
P	0.8	1.2	0.0005	0.001
K	0.8	1.2	0.005	0.010

- calculation of multiplication factor for light stress (f_{LAI})
  q_l: 3 $\quad$ LAI_{cr}: 4 $\quad$ min. f_{LAI}: 0 $\quad$ max. f_{LAI}: 3

- calculation of multiplication factor for moisture stress
  q_m: 3 $\quad$ max. f_{MS}: 3 $\quad$ min. f_{MS}: 4

Parameters for the calculations of specific leaf area (SLA)

q 1.99 $\quad$ r 405 $\quad$ s 2.51 $\quad$ t 6804

AFGEN (WOODDM, RRWFA); relative rate of wood fall (1/yr) as depending on wood dry matter (kg/ha)

WOODDM	0	5000	15000	50000	999000
RRWFA	0.27	0.27	0.05	0.01	0.01

Ratio of the fraction of coarse-root turnover to the fraction of wood fall (f_{CRTR}): 2

- Distribution constants, K_d. The distribution factor for K (K_d = adsorbed content/solution content) was set at 100 (Section 3.2.4). Linear adsorption of N and P was not considered in the present model runs (in principle the labile pool of P corresponds to adsorbed P).

- Relative rates of removal of forest floor leaf (RRFFLR) and of forest floor wood (RRFFWR) are the reciprocal values of residence time of forest floor leaves and wood (Section 3.3.1). The residence time of FFL is estimated at 1 yr. The residence time of FFW (branches + stems) was estimated at 15 yr from data by Vooren (1985) (Noij, 1988). The relative rates of removal of fine-root (RRFRDR) debris and coarse-root (RRCRDR) debris (Section 3.3.1) were set equal to those of forest floor leaf and wood, respectively.

- "Initial ages" of the organic pools are discussed in Section 3.3.2.1, 3.3.2.4, and in Part I, Section 3.3.3.1.

- The ratios DA, CN_m, CP_m and NP_m of micro-organisms and the temperature correction factor (f_t) are discussed in Part I, Section 3.2.1.

- The ratio CDM is an average of values ranging from 0.45 to 0.6 for various compounds in plant materials.

- The fractions of potassium in freshly fallen leaves (FFFLL), and in fresh fine-root debris (FFRDL) that are transported directly into solution (Section 3.3.4), are estimated. The same holds for the value f_l, the fraction of initially present nutrients leached during the time step.

- Below the minimum concentrations in soil solutions ($C_{min,i}$) no nutrient uptake will take place. See Section 3.4.2. Present values are first approximations.

- Intercepts and slopes of the equations relating $RATXY_{up}$ to $RATXY_{ss}$ and minimum and maximum values of the ratios N/P, P/K and K/N for uptake are used in Section 3.4.2 and derived in Part II, Section 4.

- In the calculation of nutrient uptake, the moisture volume fraction at which reduction of nutrient uptake occurs (Θ_{ru}) is a linear function (Equation III.63) of the moisture volume fraction at which reduction in water uptake occurs (Θ_r). At present Θ_{ru} is set equal to Θ_r. In other words: f_{ru} is set at 1 on this relation no literature data were available.

- Minimum and maximum leaf concentrations for N, P and K are presented in Part II, Section 5.

- The fraction of branches (FBRA) in newly formed wood was set at 33% (Section 3.4.3, Eq. III.).

- The parameters used for the calculation of nutrient concentrations in newly formed fine roots, branches and stems were derived from literature, as described in Part II, Section 5.

- The value for the transpiration ratio (TRR) was taken from Van Keulen and Wolf (1986).

- Leaf parameters that describe the leaf-wood redistribution as a function of leaf concentrations for P and N. The parameter in Column 1 is the minimum nutrient concentration in falling leaves, a parameter used to describe the non-linear part of the redistribution function as shown in Fig II.14 (Part II, Section 6.). The parameter in Column 2 is the slope in the linear part of the redistribution function in Fig. II. 14, and that in Column 3 is the intercept. Concentrations are expressed in kg/kg.

- The values for the reduction fraction ($FLRDU_i$) are estimated at 0.28 for dry matter and at 0.15 for K (Part II, Section 6).

- The relative rate of fine root turnover (RRFRTR), was set at 1/yr as an average value, obtained from literature study (Part II, Section 2).

- The parameters for the calculation of specific leaf area and those for the calculation of the relative rate of leaf fall were derived in Part II, Section 7.

- A tabulated function for the relative rate of woodfall (RRWFA), a so-called AFGEN function, is used to calculate the fraction of woodfall per time step (Section 3.4.4) (Noij,1988). This is done by linear interpolation between the given values for pairs of wood dry matter and relative rates of wood fall. The values of these pairs have been extracted from literature data for the Taï-area (Jaffré & de Namur (1983). For a mature forest the relative rate of woodfall is 0.01 /yr, indicating an average age of 100 years. Rather arbitrarity, the fraction of coarse-root turnover is set at two times the fraction of wood fall.

Boundary fluxes

See Table III.5.

- Annual atmospheric deposition rates ($ARDEP_i$). For phosphorus deposition no direct measurements were available for the Taï area. The given value of 0.89 kg/ha per year was derived from steady state calculations with the NUTCYC model (Noij,1988). Recent studies by Stoorvogel (1992) show that $ARDEP_i$ for

P (dry deposition) and K (wet deposition) may be set at 0.1 and 2.1 kg/ha per year, respectively. Nitrogen bulk deposition (wet + dry) was derived form data of Penning de Vries & Djitiye (1982, p. 339 and 240) and Poels (1987). The given value for N deposition is higher than found in literature; this was done to account for N fixation.

- The value of 20 kg/ha for annual K weathering ($ARWEA_K$) is a first approximation. It is high enough to prevent deficiency of K in the vegetation.

- The relative rates of erosion of all pools considered in the mineral layer ($RREROS_k$) were set at 0.001 per year. This value is based on a measured average erosion rate of 2500 kg/ha per year in the Taï area (Collinet et al., 1984), for a mineral layer of 20 cm thickness and a bulk density of 1350 kg/m^3. The relative erosion rates of forest floor leaves and wood are estimated at 15% per year. This fraction is an average of severe erosion in periods with high precipitation and very little erosion in dry periods. Only the part of the forest floor leaves and wood that is susceptible to decomposition (FFFLR and FFFWR) is susceptible to erosion.

Table III.5 Boundary fluxes and boundary flux regulating variables

Annual rates of atmospheric deposition ($ARDEP_i$) (kg/ha per year)

N	30.0	P	0.897	K	4.0

Annual rate of weathering

$ARWEA_K$: 20 (kg/ha per year)
Relative rates of erosion of pool k ($RREROS_k$) (1/year)

Pool	$RREROS_k$	Pool	$RREROS_k$	Pool	$RREROS_k$
ININ	0.001	ORST	0.001	FFL	0.15
INST	0.001	ORML	0.001	FFW	0.15
INLAP	0.001	ORLA	0.001	CRD	0.001
				FRD	0.001

5 SENSITIVITY ANALYSIS OF THE MOISTURE CYCLING SUB-MODEL

5.1 Introduction

If we assume the model structure, process formulations and numerical procedures to be correct, we can analyse the sensitivity and the uncertainty of the model output. The sensitivity of a model output parameter (y) for a model input parameter (x) can be defined as the relative change of y with x, when x is changed by a small fixed amount; e.g 5%. The uncertainty of y due to the uncertainty in x can be defined as the relative increase of uncertainty in y due to a given uncertainty in x (Janssen et al., 1990b). The objective of sensitivity analysis is to get a quantitative overview of the relative importance of the input parameters. In contrast to sensitivity analysis, uncertainty analysis requires knowledge about the actual parameter uncertainty. The interpretations of results of a sensitivity analysis and an uncertainty analysis are strongly related. For example, the model may be extremely sensitive to the amount of precipitation, but still this parameter may be excluded in an uncertainty analysis because it can be measured very accurately.
The sensitivity of the moisture cycling sub-model was examined by the parameter perturbation method. Although essential for evaluation of model performance, an uncertainty analysis of DYNAMITE has not yet been carried out. Time was not available, and information about actual input uncertainties was inadequate.

The sensitivity of the complete model was examined by a Monte Carlo technique. A former version of the model was used. Due to some unfortunately chosen boundary conditions, however, the calculated results were far from realistic. After that analysis and during the final editing of this publication, in both sub-models some parts were adjusted and refined. No time was available to repeat the Monte Carlo analysis with the model in its stage at the end of the project. Therefore only results of the sensitivity analysis of the moisture cycling sub-module are reported.

5.2 Subjects and procedure

An independent sensitivity analysis was carried out on the moisture cycling sub-model of DYNAMITE. The sensitivity of transpiration, percolation and capillary rise (second soil layer) was examined for the following parameters:

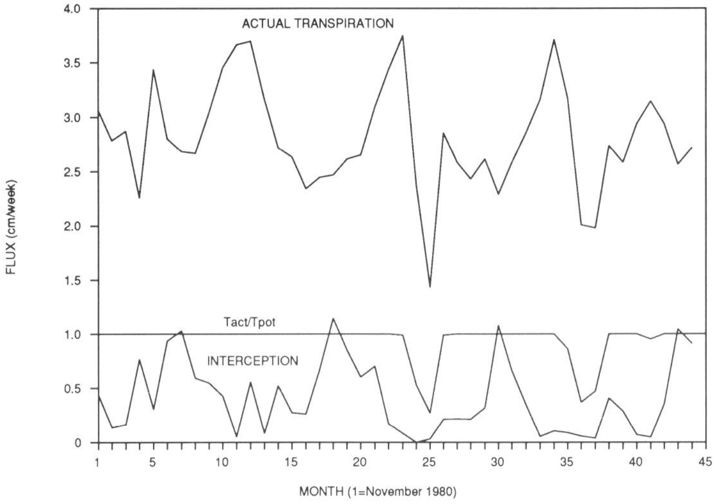

Fig. III.9 DYNAMITE simulations of actual transpiration, the ratio of actual over potential transpiration and canopy interception for the Kabo experimental forest, Suriname, between November 1980 and May 1984.

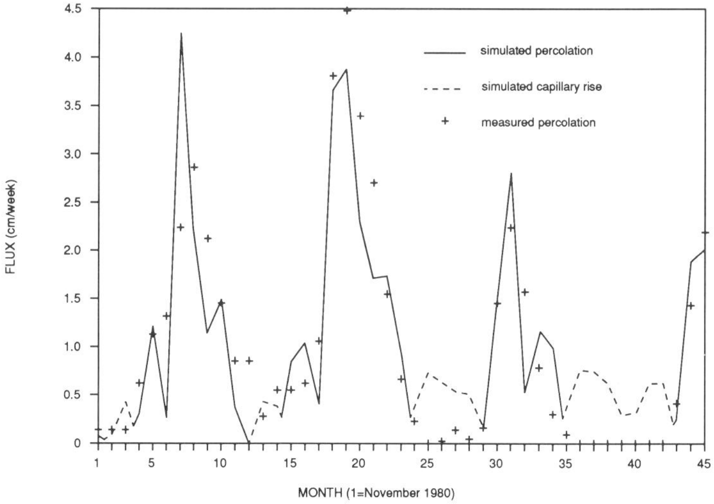

Fig. III.10 DYNAMITE simulations of soil water fluxes at 450 cm depth for the Kabo experimental forest, Suriname, between November 1980 and May 1984.

- the maximum percolation rate J_{max};
- the rainfall amount at varying J_{max};
- the rainfall distribution at varying J_{max};
- the time step for simulation.

J_{max} is highly uncertain, because J_{max} is an empirical parameter, which bares no unique relationship with the hydraulic conductivity and thus can only be determined by calibration. The uncertainty in the water balance calculations due to uncertainty or poor calibration of J_{max} is closely related to the amount and the distribution of rainfall, and the simulation time step. Although these additional parameters generally are not important sources of uncertainty for a specific simulation study, they still are included in the sensitivity analysis, to get a more general impression of model performance.

The tests were carried out with observed biweekly cumulative rainfall and pan evaporation data from Kabo, Suriname (Poels, 1987), over a period of 3.7 years, starting in November, 1980. In the water balance calculations two soil layers were considered; a top layer of 500 mm thickness, and a bottom layer of 3700 mm thickness. Potential water uptake in the top layer was 80% and in the bottom layer 20% of potential transpiration. Capillary rise was considered only for the bottom layer. Initial water contents in the simulations were set at field capacity.
To give an impression of the performance of the water balance model, simulated actual transpiration and interception are shown in Fig. III.9 and simulated and measured percolation at 4.5 m depth in Fig. III.10. The rare occurrence of transpiration reduction reflects the ample availability of rainfall (Fig. III.9). The fair agreement between simulated and observed monthly percolation illustrates the applicability of the moisture cycling sub-model (Fig. III.10).

To examine the effect of rainfall amount on the water balance, simulations were carried out with 50% and 25% of the original biweekly rainfall amounts in Kabo. To examine the effect of rainfall distribution on the water balance, the original amount of monthly rainfall was assigned to the first two weeks, while the second two weeks were without rain. In both tests pan evaporation was not modified.
To examine the effect of the time step, simulations were carried with a time step of two weeks and one month. For this purpose biweekly rainfall and pan evaporation data were added up to monthly data.

5.3 Results, discussion and conclusions

Maximum percolation rate

The slopes of the curves in Fig. III.11 are a measure of the sensitivity to the water balance term J_{max}. Transpiration, percolation and capillary rise are highly sensitive to J_{max} at low J_{max} values and unsensitive at high J_{max}-values. At low J_{max} values,

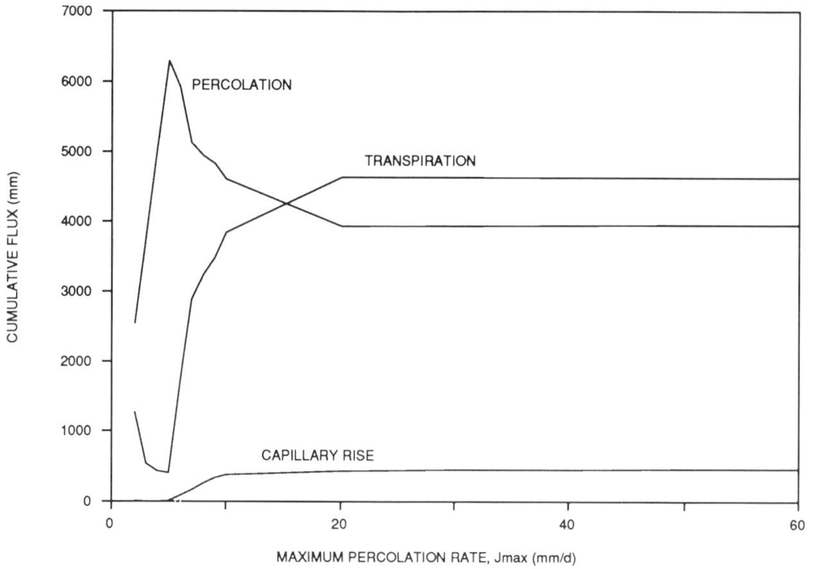

Fig. III.11 Sensitivity of DYNAMITE simulations of transpiration, capillary rise and percolation to uncertainty in the maximum percolation rate (J_{max}) using hydrologic input data for Kabo, Suriname.

transpiration is reduced because of oxygen shortage. For the theoretical case where $J_{max} = 0$, there will be no transpiration and percolation. The occurrence of a transpiration minimum, and consequently a percolation maximum, at low J_{max} is caused by counteraction of the effects of oxygen shortage (Equations III.14 and III.15) and more efficient water conservation. A short-lived transpiration reduction due to oxygen stress may or may not be compensated by higher soil water availability in a following dry period. The critical J_{max} value is about 5 mm/d. For $J_{max} > 5$ mm/d, oxygen stress will occur less frequently and hence transpiration steadily increases till it reaches a plateau value for $J_{max} > 20$ mm/d.

Maximum percolation rate and rainfall amount

Independent of the absolute values of the fluxes, insensitivity to J_{max} is reached at high values of J_{max}. With increasing rainfall, the sensitivity of percolation, transpiration and capillary rise to J_{max} increases (Fig. III.12a,b,c). Increased sensitivity is caused by more frequent occurrence of waterlogging and subsequent transpiration reduction, with increasing rainfall. Transpiration and percolation increase, while capillary rise decreases with increasing rainfall. The sensitivity of percolation to the amount of rainfall is higher than that of transpiration, because transpiration is limited by the transpirative demand, while percolation is limited by the amount of rainfall. In the simulation with 25% of the original rainfall, no percolation occurs, indicating that all rainfall is consumed by evaporation and transpiration.

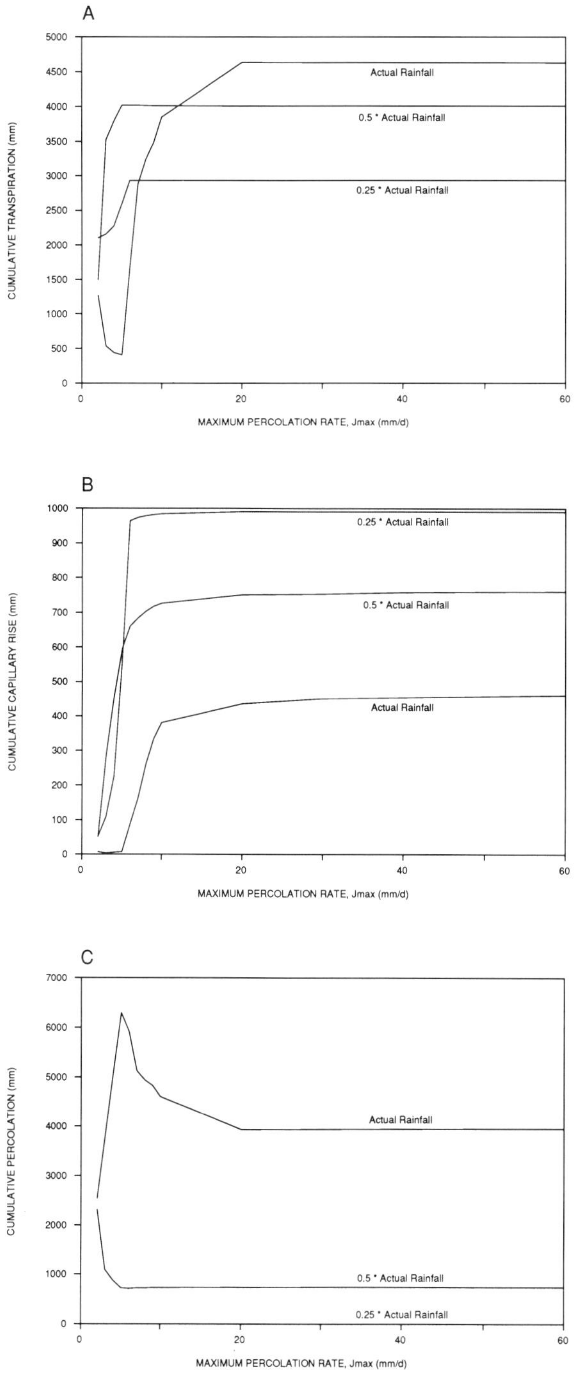

Fig. III.12 Sensitivity of DYNAMITE simulations of (A) transpiration, (B) capillary rise and (C) percolation to uncertainty in the maximum percolation rate (J_{max}) at 50% and 25% of the original amount of rainfall.

Maximum percolation rate and rainfall distribution

Percolation as well as its sensitivity to J_{max} slightly increase with a less uniform rainfall distribution (Fig. III.13). Apparently, the water storage in the profile is sufficient to satisfy the transpirative demand of the vegetation also during the imposed intermittent dry periods of Case B. It may be expected that percolation, and its sensitivity to J_{max} will increase with decreasing uniformity of the rainfall distribution.

Maximum percolation and time step

Decreasing the time step from four to two weeks has no important effect on simulated transpiration, percolation and capillary rise, and their sensitivity to J_{max} (Fig. III.14). Decreasing of the time step has the largest effect on percolation, indicating a somewhat more efficient water use with smaller time steps.

Conclusions

The water balance calculations are not sensitive to J_{max} above J_{max} values of 20 mm/d in case of high amounts of well distributed rainfall. Such a situation is typical in the wet tropics. The sensitivity of the water balance to uncertainty in J_{max}, shows the largest increase when the amount of rainfall is lowered. Extra attention should be paid to calibration of the water balance model, when applying DYNAMITE to forest sites in dry tropical regions or to sites with poorly permeable soil layers. A maximum time step for realistic simulation of reduced soil water uptake by plants is four weeks.

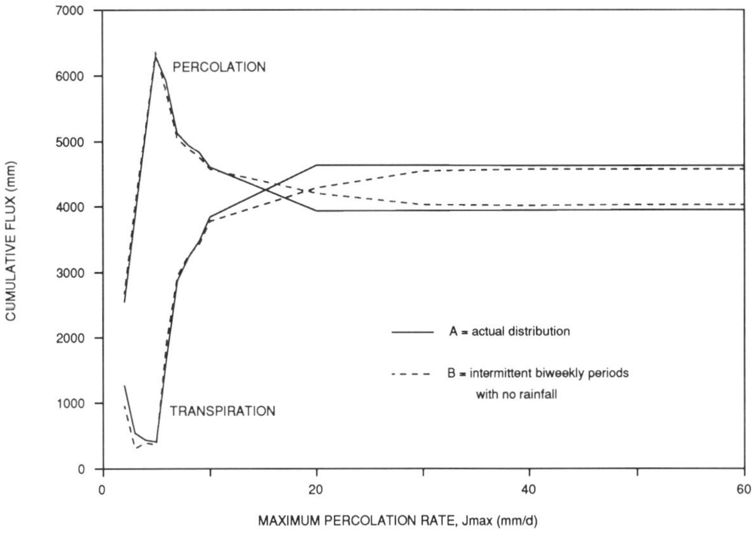

Fig. III.13 Sensitivity of DYNAMITE simulations of transpiration and percolation to uncertainty in the maximum percolation rate (J_{max}) for the original rainfall distribution and for a hypothetical distribution with alternate two-week periods with and without rainfall.

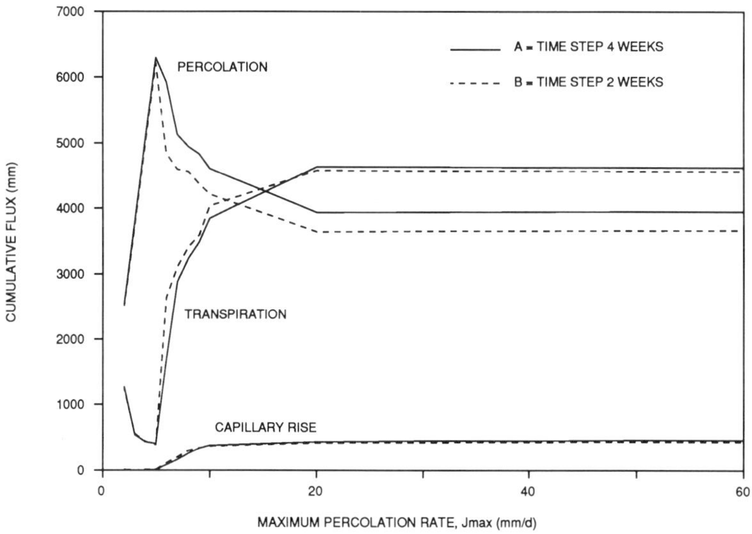

Fig. III.14 Sensitivity of DYNAMITE simulations of transpiration, capillary rise and percolation to uncertainty in the maximum percolation rate (J_{max}) for time steps of two and four weeks.

6 FUTURE MODIFICATION

Modifications of DYNAMITE to be considered in the future are of two kinds:
- improvement of the present model
- extension of the present model

Examples of modifications of the first kind refer to:
- reduction of nutrient uptake due to water or oxygen shortage;
- the distribution of absorbed nutrients among the various plant components;
- turnover rate of fine roots;
- translocation fraction of fine to coarse roots;
- mortification rates of coarse roots and wood;
- weathering of inorganic stable P-pools and adsorption/desorption model for inorganic labile P;
- conversion rate of stable organic pool.

Examples of modifications of the second kind refer to:
- the number of elements. In addition to P, N, K and C, also Ca, Mg, H, Al may be included. In order to simulate a complete solute balance also Na, Cl and SO4 should be added. Furthermore a distinction between NH4 and NO3 may be made;
- the number of soil layers;
- lateral transport; distinction of soil segments for simulation a sequence of soil profiles on a slope;
- the effect of pH and soil moisture on mineralization coefficients;
- a soil heat balance, to allow automatic correction of some of the rate coefficients for seasonal variation and for different geographic regions. As a first approach the temperature wave at the surface could be considered sinusoidical. Simple analytical procedures are available to predict the phase shift and amplitude dampening of the temperature wave in the soil;
- N fixation. A considerable proportion of the wood species in tropical forest belongs to the leguminous group (up to 30%). Many of them are living in symbiosis with N-fixing bacteria. Especially in cases where soil N is low, N fixation should be considered. The least complicated approach is assumption of a fixed N-fixation flux. A dynamical approach would entail a separate simulation of the vegetation contributing to N fixation. Up to now we did not distinguish between different tree species;
- denitrification which is an important loss of N in soils which are wet during part of the year;
- effects of pH and Al on root growth; such a relationship requires an independent simulation of pH and Al as mentioned above;

- dust deposition of K in combination with a first order weathering rate equation. In poor sandy soils the mineral pool of K will be limited. Depletion of the mineral K pool by weathering and resupply of this pool by atmospheric deposition will affect the K availability to plant growth, in particular for long-term simulation runs;
- a separate routine for easy and automatic scenario analysis. Such a routine will allow adaption of fluxes and pools at any point of time during a simulation. With the help of such a routine the effects of timber felling, burning and cultivation of agricultural crops can be evaluated rather easily. Some examples of system characteristics subject to changes after such activities are the following:
  Timber felling:
  - decreased soil cover;
  - decreased transpiration and interception but increased soil evaporation;
  - transfer of living biomass to litter and root debris;
  - increased erosion;
  - increased soil temperature which will enhance decomposition processes.

  Burning:
  - transition of nutrients stored in organic pools to inorganic pools;
  - increase of pH which will change the availability of nutrients to the plant.

  Cultivation of agricultural crops:
  - changed plant uptake and growth characteristics;
  - removal of nutrients and organic matter due to harvesting;
  - changed erosion fraction;
  - fertilizer addition;
  - changed soil hydraulic properties due to tillage.

REFERENCES

Bultot, F., Dupriez, G.L. & Bodeux, A. (1976). Interception potentielle journalière de la pluie par divers peuplement forestier. J. Hydrol. **31**: 381-392.

Casenave, A., Flory, J., Guiguen, N., Ranc, N., Simon, J.M., Toilliez, J. & Tourne, M. (1980). Étude hydrologique des bassins de Taï. Campagnes 1978-1979. Rapport ORSTOM multigr. ORSTOM, Paris, France.

Collinet, J., Monteny, B. & Pouyand, B. (1984). Le milieu physique. in: Guillaumet, J.L., Couturier, G. et Dosso, H. (eds). Recherche et aménagement en milieu forestier tropical humide: le project Taï de Côte d'Ivoire. Notes technique du MAB **15**: 35-58. Unesco, Paris, France.

Driessen, P. (1986). The water balance of the soil. in: Van Keulen, H. & Wolf. J. (eds.). Modelling of agricultural production: weather, soils & crops. Simulation Monographs: 76-116. PUDOC, Wageningen, The Netherlands.

Feddes, R.A., Kowalik, P.J. & Zaradny, H. (1978). Simulation of field water & crop yield. Simulation Monographs. PUDOC, Wageningen, The Netherlands.

Fritsch, E. (1982). Evolution des sols sous recru forestier apres mise en culture traditionelle dans le Sud-Ouest de la Côte d'Ivoire. Rapport ORSTOM multigr. ORSTOM, Paris, France.

Goldstein, R.A., Geherini, S.A., Chen, C.W., Mak, L., & Hudson, R.J.M. (1984). Integrated lake acidification study (ILWAS): a mechanistic ecosystem analysis. Philos. Trans. Soc. London, Series B **305**: 405-425

Guillaumet, J.L., Couturier, G. & Dosso, H. (1984). Recherche et aménagement en milieu forestier humide: le projet Taï de Côte d'Ivoire. Notes techniques du MAB **15**. Unesco, Paris, France.

Huttel, C. (1977). Etude de quelques characteristiques structurales de la végétation de bassin versant de l'Andrenisrou, Rapport ORSTOM multigr. ORSTOM, Paris, France.

Jackson, I.J. (1975). Relationships between rainfall parameters and interception by tropical forests. J. Hydrol. **24**: 215-238.

Jaffré, T. & de Namur, Ch. (1983). Evolution de la biomasse végétale epigée au cours de la succession secondaire dans le Sud-Ouest de la Côte d'Ivoire. Acta Oecologica Oecol. Plant. **4**: 250-272.

Janssen, B.H. (1984). A simple method for calculating decomposition and accumulation of "young" organic matter. Plant and Soil **76**: 297- 304.

Janssen, B.H. (1986). Een één-parameter model voor de berekening van de decompositie van organisch materiaal. Vakbl. Biol. **66**: 433- 436.

Janssen, B.H. & Noij, I.G.A.M. (1993). A simple model and a linear equation for the calculation of nitrogen and phosphorus mineralization. Plant and Soil.

Janssen, B.H., Noij, I.G.A.M., Wesselink, L.G. & Van Grinsven, J.J.M. (1990a). Simulation of the dynamics of nutrients and moisture in tropical ecosystems. Fertilizer Research **26**: 145-156.

Janssen, P.H.M., Slob, W. & Rotmans, J. (1990b). Gevoeligheidsanalyse en onzekerheidsanalyse: Een inventarisatie van ideeën, methoden en technieken. RIVM rapport 958805001, Bilthoven, The Netherlands.

Jenkinson, D.S. & Ayanaba, A. (1977). Decomposition of carbon-14 labeled plant material under tropical conditions. Soil Sci. Soc. Amer. J. **41**: 912-915.

Noij, I.G.A.M. (1988). Een eenvoudig simulatiemodel voor nutriëntenbeperkte fytomassa produktie in tropische bosgebieden: NUTCYC. Verslagen en mededelingen 1988-1. Department of Soil Science and Plant Nutrition, Agricultural University, Wageningen, The Netherlands.

Penman, M.L. (1948). Natural evaporation from open water, bare soil and grass. Proc. R. Soc. London **193**: 120-145.

Penning de Vries, F.W.T. & Djiteye, M.A. (1982). La productivité de pâturages saheliens: une étude des sols, des végétations et de l'exploration de cette resource naturelle. Agric. Res. Report **918**. PUDOC, Wageningen, The Netherlands.

Poels, R.L.H. (1987). Soils, water and nutrients in a forest ecosystem in Suriname. Ecology and management of tropical rain forests in Suriname. Agricultural University, Wageningen, The Netherlands.

Press, W.A., Flannery, B.P., Teukolsky, S.A. & Vetterling, W.T. (1986). Numerical Recipes. Cambridge Univ. Press, United Kingdom.

Singh, P. & Sceicz, G. (1979). The effect of intercepted rainfall on the water balance of a hardwood forest. Water Resour. Res. **15**: 131-138.

Staritsky, I.G. (1988). Een model voor de nutriëntencyclus in het regenwoud in Sarawak. (Internal report). Department of Soil Science and Plant Nutrition, Agricultural University, Wageningen, The Netherlands.

Stoorvogel, J.J. (in prep.). Gross inputs and outputs of nutrients in undisturbed forest, Taï Area, Côte d'Ivoire.

Stumm, W., Furrer, W., Wieland, E. & Zinder, B. (1985). The effect of complex-forming ligands on the dissolution of oxides and aluminosilicates. in: Drever, J.I. (ed.). The chemistry of weathering: 55-73. D. Reidel Publ. Co., Dordrecht, The Netherlands.

Thornthwaite, C.W. & Holzman, B. (1939). The determination of evaporation from land and water surfaces. Mon. Weather Rev. **67**: 4-11.

Van Keulen, H. & Wolf. J. (eds). (1986). Modelling of agricultural production: weather, soils and crops. Simulation Monographs. PUDOC, Wageningen, The Netherlands.

Van Roestel, J. (1984). Transpiratie en interceptie van bos: een literatuurstudie. Bos en Water 7B. Studiecommissie Waterbeheer Natuur, Bos en Landschap, Utrecht, The Netherlands.

Vooren, A.P. (1985). Patterns in tree and branch-fall in a West-African rain forest. Report No. D 85-05. Department of Silviculture, Agricultural University, Wageningen, The Netherlands.

Wolf, J., de Wit, C.T., Janssen, B.H. & Lathwell, D.J. (1987). Modeling long-term crop response to fertilizer phosphorus: I. The model. Agronomy J. **79**: 445-451.

APPENDIX III.1 LIST OF SYMBOLS IN ALPHABETICAL ORDER

Regression constants without a physical meaning are not included in the list.

Symbol	Description	unit
ADS_i	amount of adsorbed substance i	kg/ha
$ADS_{i,b}$	amount of adsorbed substance i at the beginning of the time step	kg/ha
$ADS_{i,e}$	amount of adsorbed substance i at the end of the time step	kg/ha
AGFEN (X, Y)	linear interpolation function	
AME_{pan}	average monthly pan evaporation	mm
AMR	average monthly rainfall	mm
$ARDEP_i$	annual rate of atmospheric deposition of substance i	kg/ha per year
$ARWEA_i$	annual rate of weathering of kg/ha per year substance i	kg/ha per year
$BRANCA_i$	maximum concentration (mass fraction) of substance i in branches	kg/kg
$BRNC_i$	concentration (mass fraction) of substance i in newly formed branches	kg/kg
C_0	carbon content of an organic poot at the beginning of the time step	kg/ha
C_{dis}	first order rate constant for dissimilation	1/yr
$C_{dis,k}$	C_{dis} of pool k	1/yr
$C_{min,i}$	minimum soil solution concentration required for uptake of nutrient i	kg/m^3
CDM	ratio of carbon to dry matter	kg/kg
CN_m	C-N ratio of micro-organisms	kg/kg
CN_s	C-N ratio of substrate (= organic pool)	kg/kg
CP_m	C-P ratio of micro-organisms	kg/kg
CP_s	C-P ratio of substrate (= organic pool)	kg/kg
CR	coarse roots	--
CRD	coarse-root debris	--
CRD_i	amount of substance i present in coarse-root debris	kg/ha
$CRNC_i$	concentration (mass fraction) of substance i in newly formed coarse roots	kg/kg
$CROOT_i$	total amount of substance i present in coarse roots	kg/ha

$CROOTC_i$	concentration (mass fraction) of nutrient i in coarse roots	kg/kg
$CRTCA_i$	maximum concentration (mass fraction) of substance i in coarse roots (A = accumulation)	kg/kg
$CRTCD_i$	minimum concentration (mass fraction) of substance i in coarse roots (D = dilution)	kg/kg
D	thickness of soil layer	mm
DA	dissimilation-assimilation ratio of micro-organisms	--
DAYSWR	average number of days with rain per month	--
E	evaporation	mm/time step
E_0	open-water evaporation	mm/time step
E_{act}	actual soil evaporation	mm/time step
E_i	direct evaporation from the cacopy	mm/time step
E_{pan}	pan evaporation	mm/time step
ET	evapotranspiration	mm/time step
ET_{pot}	potential evapotranspiration	mm/time step
ET_{pot}'	evaporative demand of soil and vegetation	mm/time step
$F_{dep,i}$	flux of subsytance i by atmospheric wet deposition	kg/ha per time step
$F_{in,i}$	flux of substance i from overlying soil layer	kg/ha per time step
$F_{min,i}$	flux of substance i by mineralization from all organic pools	kg/ha per time step
$F_{out,i}$	flux of substance i by leaching	kg/ha per time step
$F_{wea,i}$	flux of substance i by weathering from all inorganic pools	kg/ha per time step
FBRA	fraction of wood allocated to branches	kg/kg
FCRDR	fraction of coarse-root debris that is removed per time step	1/time step
FCRTR	fraction of coarse roots that is turned over per time step	1/time step
FER_k	erosion fraction of pool k per time step	1/time step
$FFFILL_K$	fraction of K in fresh forest floor leaves that is leached immediately	kg/kg
FFFLR	fraction of forest floor leaves that is removed per time step	1/time step
FFFWR	fraction of forest floor wood that is removed per time step	1/time step
FFL	forest floor leaves (leaf litter)	
FFL_i	total amount of substance i present in forest floor leaves	kg/ha

FFRCRT	fraction of rate of fine-root turnover that is translocated to coarse root	kg/kg
$FFRDL_K$	fraction of K in fresh fine-root debris that is leached immediately	kg/kg
FFRDR	fraction of fine-root debris that is removed per time step	1/time step
FFRTR	fraction of fine roots that is turned over per time step	1/time step
FFW	forest floor wood (wood litter)	
FFW_i	total amount of substance i present in forest floor wood	kg/ha
FLFA	fraction of leaves that falls	1/time step
$FLRDU_i$	reduction fraction of total amount of substance i in leaves before leaf fall	kg/kg
$FRNC_i$	concentration (mass fraction) of substance i in newly formed fine roots	kg/kg
$FRNC_{il}$	concentration (mass fraction) of the limiting nutrient in newly formed fine roots	kg/kg
$FROOT_i$	total amount of substance i present in all fine roots	kg/ha
$FROOTC_i$	concentration (mass fraction) of substance i in fine roots	kg/kg
$FROOTCA_i$	maximum concentration (mass fraction) of substance i in fine roots	kg/kg
$FROOTCD_i$	minimum concentration (mass fraction) of substance i in fine roots	kg/kg
FWFA	fraction of wood that falls per time step	1/time step
FWU_j	potential water uptake from soil layer j as fraction of total potential water uptake (= potential transpiration)	mm/mm
GROW	growth (dry matter) per time step	kg/ha per time step
GROWCR	growth of coarse roots	kg/ha per time step
GROWFR	growth of fine roots	kg/ha per time step
GROWL	growth of leaves	kg/ha per time step
$GROWL_{max}$	maximum leaf growth as determined by T_{act}	kg/ha per time step
$GROWLA_i$	growth of leaves when the concentration (mass fraction) of nutrient i would be maximum	kg/ha per time step

$GROWLA_{max}$	maximum of $GROWLA_i$ for i = N, P or K	kg/ha per time step
$GROWLD_i$	growth of leaves when the concentration (mass fraction) of nutrient i would be minimum	kg/ha per time step
$GROWLD_{min}$	minimum of $GROWLD_i$ for i = N, P or K	kg/ha per time step
GROWW	growth of wood (stem + branches)	kg/ha per time step
I_a	canopy interception of rain	mm/day, or mm/time step
ININP	phosphorus in inorganic inert pool	kg/ha
INLAP	phosphorus in inorganic labile pool	kg/ha
INSTP	phosphorus in inorganic stable pool	kg/ha
J_{ass}	assimilation flux	kg/ha per time step
$J_{ass,i}$	J_{ass} of substance i	kg/ha per time step
J_{conv}	conversion flux	kg/ha per time step
$J_{conv,i}$	J_{conv} of substance i	kg/ha per time step
J_{cr}	maximum capillary rise	mm/day, or mm/time step
J_{dat}	$J_{dis} + J_{tra}$	kg/ha per time step
$J_{dat,i}$	J_{dat} of substance i	kg/ha per time step
J_{dis}	dissimilation flux	kg/ha per time step
$J_{dis,k}$	J_{dis} of pool k	kg/ha per time step
J_{in}	incoming water flux (in soil layer)	mm/day, or mm/time step
$J_{lsp,k}$	flux of instantaneous release of K from the last sub-pool of a primary organic pool at the end of the time step	kg/ha per time step
J_{max}	maximum percolation rate	mm/day
J_{min}	mineralization flux	kg/ha per time step
$J_{min,i}$	J_{min} of substance i	kg/ha per time step
J_{out}	outgoing water flux (from soil layer)	mm/day, or mm/time step

$J_{out,\Theta}$	outgoing water flux when the water volume fraction of the soil layer is Θ	mm/day, or mm/time step
$J_{rel,K}$	release flux of K from primary organic pools	kg/ha per time step
J_{tra}	transfer flux	kg/ha per time step
$J_{tra,i}$	J_{tra} of substance i	kg/ha per time step
J_{Θ}	water flux when water volume fraction is Θ	mm/day, or mm/time step
K_d	distribution coefficient of adsorbed K and K in soil solution	(kg/ha)/(kg/ha)
LAI	leaf area index	ha/ha
LEAF	leaves	
$LEAF_i$	total amount of substance i present in leaves	kg/ha
$LEAFC_i$	concentration (mass fraction) of substance i in living leaves	kg/kg
$LEAFC_{i,b}$	concentration (mass fraction) of substance i at the beginning of the time step	kg/kg
$LEAFCA_i$	maximum concentration (mass fraction) of substance i in living leaves	kg/kg
$LEAFCD_i$	minimum concentration (mass fraction) of substance i in living leaves	kg/kg
$LEAFLC_i$	lowest possible concentration (mass fraction) of substance i in falling leaves	kg/kg
LNC_i	concentration (mass fraction) of substance i in newly formed leaves	kg/kg
M	actual mineral mass	kg/ha
M_0	initial mineral mass	kg/ha
ME_{pan}	monthly pan evaporation	mm per month
MR	monthly rain fall	mm per month
NAN_i	amount of nutrient i in soil solution that is not available to plants	kg/ha
$NGROWCR_i$	net increase of amount of substance i in coarse roots	kg/ha per time step
$NGROWFR_i$	net increase of amount of substance i in fine roots	kg/ha per time step
$NGROWL_i$	net increase of amount of substance i in leaves	kg/ha per time step
$NGROWW_i$	net increase of amount of substance i in wood (stem + branches)	kg/ha per time step
ORLA	organic labile pool	--
$ORLA_i$	total amount of substance i present in ORLA	kg/ha

ORML	organic moderately labile pool	--
$ORML_i$	total amount of substance i present in ORML	kg/ha
ORST	organic stable pool	--
$ORST_i$	total amount of substance i present in ORST	kg/ha
P	daily precipitation	mm/day
$POOL_{k,i}$	amount of substance i in pool k	kg/ha
PORP	primary organic pool (= FFL, FFW, FRD, or CRD)	--
$POPR_i$	amount of substance i in primary organic pool	kg/ha
Q	average precipitation intensity on days with rain during actual time step in the model	mm/day
Q_{av}	average precipitation intensity on days with rain during the whole simulation period	mm/day
$RATX,Y_{ss}$	ratio of amount of nutrient X to amount of nutrient Y in soil solution	--
$RATX,Y_{up}$	ratio of uptake of nutrient X to uptake of nutrient Y	--
$RATX,Y_{up,max}$	maximum value that can be obtained for $RATX,Y_{up}$	--
$RATX,Y_{up,min}$	minimum value that can be obtained for $RATX,Y_{up}$	--
RCRTR	rate of coarse-root turnover	kg/ha per time step
$RCRTR_i$	rate of transfer of substance i from coarse roots to coarse-root debris	kg/ha per time step
$RDEP_i$	rate of atmospheric deposition of substance i	kg/ha per time step
RFRCR	rate of transfer from fine roots to coarse roots	kg/ha per time step
$RENW_i$	rate of wood entering by nutrient i	kg/ha per time step
$REROS_k$	rate of erosion of pool k	kg/ha per time step
$RFRCR_i$	rate of transfer of substance i from fine roots to coarse roots	kg/ha per time step
RFRSL	rate of fine-root sloughing	kg/ha per time step
$RFRSL_i$	rate of transfer of substance i from fine roots to fine-root debris	kg/ha per time step
RFRTR	rate of fine-root turnover	kg/ha per time step
RLFA	rate of leaf fall	kg/ha per time step

$RLFA_i$	transfer of substance i by leaf fall	kg/ha per time step
$RILIS_P$	rate of P transfer from inorganic labile to inorganic stable pool	kg/ha per time step
$RILSS_P$	rate of P transfer from inorganic labile pool to soil solution	kg/ha per time step
$RISIL_P$	rate of P transfer from inorganic stable to inorganic labile pool	kg/ha per time step
$RLWRE_i$	rate of redistribution of nutrient i from leaf to wood	kg/ha per time step
RRCRDR	relative rate of removal of coarse-root debris	1/yr
RRCRTR	relative rate of coarse-root turn over	1/yr
$RREROS_k$	relative rate of erosion of pool k	1/yr
RRFFLR	relative rate of removal of forest-floor leaves	1/yr
RRFFWR	relative rate of removal of forest-floor wood	1/yr
RRFRDR	relative rate of removal of fine-root debris	1/yr
RRFRTR	relative rate of fine-root turn over	1/yr
RRILIS	relative rate of transfer from inorganic labile to inorganic stable pool	1/yr
RRILSS	relative rate of transfer from inorganic labile pool to soil solution	1/yr
RRISIL	relative rate of transfer from inorganic stable to inorganic labile pool	1/yr
$RRLF_i$	relative rate of leaf fall as determined by substance i only	1/yr
RRLFA	relative rate of leaf fall	1/yr
RRWFA	relative rate of wood fall	1/yr
RWFA	rate of wood fall	kg/ha per time step
$RWFA_i$	transfer of substance i by wood fall	kg/ha per time step
S	sink term; ratio of actual and potential water uptake from soil layer j	
SFI	soil fertility index, ratio of actual leaf growth and maximum leaf growth ($GROWL/GROWL_{max}$)	--
SLA	specific leaf area	m^2/kg
SSOL	soil solution	--
$SSOL_i$	amount of substance i in soil solution	kg/ha
$SSOL_{i,b}$	amount of substance i in soil solution atthe beginning of the time step	kg/ha
$SSOL_{i,e}$	amount of substance i in soil solution at the end of the time step	kg/ha

$STEMCA_i$	maximum concentration (mass fraction) of substance i in stems	kg/kg
$STEMNC_i$	concentration (mass fraction) of substance i in newly formed stems	kg/kg
$SUPCR_i$	amount of nutrient i that cannot be retained in coarse roots, and is sent back to solution	kg/ha per time step
$SUPW_i$	amount of nutrient i that is not retained in newly formed wood, and is sent back to soil solution	kg/ha per time step
T	temperature	^oC
T_{act}	actual transpiration	mm/per time step
T_{pot}	potential transpiration	mm/per time step
TINP	total inorganic P (= INLAP + INSTP + ININP)	kg/ha
TRR	transpiration ratio, kg water transpired per kg dry matter produced	kg/kg
$UP_{act,i}$	total actual uptake of nutrient i	kg/ha per time step
$UP_{act,il}$	total actual uptake of nutrient limiting fine root growth	kg/ha per time step
$UP_{max,i}$	maximum uptake of nutrient i	kg/ha per time step
$UPFR_i$	uptake of nutrient i in fine roots	kg/ha per time step
UPL_i	uptake of nutrient in leaves (difference between $UP_{i,act}$ and $UPFR_i$)	kg/ha per time step
$UPL_{act,i}$	actual uptake of nutrient i in leaves (minimum of UPL_i and $UPLA_i$	kg/ha per time step
$UPLA_i$	uptake of nutrient i in leaves at maximum concentration of i newly formed leaves	kg/ha per time step
UPW_i	direct uptake of nutrient i in wood	kg/ha per time step
W_{act}	actual uptake of water	mm/per time step
$W_{act,j}$	actual uptake of water from soil layer j	mm/per time step
$W_{pot,j}$	potential uptake of water from soil layer j	mm/per time step
WOOD	wood (stems plus branches)	
$WOOD_i$	amount of substance i in wood	kg/ha
$WOODC_i$	concentrtion (mass fraction) of substance i in wood	kg/kg

$WOODNC_i$	concentration (mass fraction) of substance i in newly formed wood	kg/kg
Y_0	initial amount of organic matter	kg/ha
Y_t	amount of organic matter remaining at time t	kg/ha
Y_{tot}	organic matter in all sub-pools of a primary organic pool	kg/ha
$Y_{tot,b}$	Y_{tot} at the beginning of the time step	kg/ha
$Y_{tot,e}$	Y_{tot} at the end of the time step	kg/ha

a	apparent initial age of organic materials	yr
$a_{i,b}$	age of organic sub-pool i at the beginning of the time step	yr
$a_{i,e}$	age of organic sub-pool i at the end of the time step	yr
c_1	correction term for precipitation in the calculation of I_a from P	--
c_2	correction term for evaporative demand in the calculation of I_a from P	--
f_1	crop factor (E_{pot}'/E_{pan})	--
f_2	fraction of energy for E_i that is obtained directly from irradiation	--
f_l	fraction of nutrients present at the beginning of a timestep, that is leached from soil solution during time step	--
f_{LAI}	correction factor for light stress in the calculation of RRLFA	--
f_{MS}	correction factor for moisture stress in the calculation of RRLFA	--
f_{ru}	ratio of water volume fraction below which reduction in nutrient uptake starts to water volume fraction below which reduction in water uptake starts	--
f_t	temperature correction factor for relative rate of decomposition organic matter	--
q_l	maximum value of f_{LAI}	--
q_m	maximum value of f_{MS}	--
q_r	ratio of $CROOT_{DM}/FROOT_{DM}$ to SFI	--
t	time	yr

α	multiplication factor in the calculation of J_Θ, to take into account Θ of underlying layer	--
β	exponent for sensitivity in the calculation of α	--
δt	time step	yr
Θ	volume fraction of water in soil	mm^3/mm^3
Θ_{ad}	volume fraction of soil water after air drying	mm^3/mm^3
Θ_b	volume fraction of water at the beginning of the time step	mm^3/mm^3
Θ_e	volume fraction of water at the end of the time step	mm^3/mm^3
Θ_{fc}	volume fraction of soil water at field capacity	mm^3/mm^3
Θ_r	volume fraction of soil water below which uptake of water is reduced due to drought	mm^3/mm^3
Θ_{ru}	volume fraction of soil water below which the uptake of nutrients is reduced	mm^3/mm^3
Θ_{wp}	volume fraction of soil water at wilting point	mm^3/mm^3
σ	discontinuous function of transpirative demand	--
$\Sigma\ in_i$	sum of net inputs of nutrient i into soil solution	kg/ha per time step
$\Sigma\ in_{k,i}$	sum of inputs of nutrient i into pool k	kg/ha per time step
$\Sigma\ out_{k,i}$	sum of output of nutrient i from pool k	kg/ha per time step

APPENDIX III. 2 MODEL USE

Design, editing, compilation and linking in TURBO PASCAL 4.0

The water balance sub-model in DYNAMITE uses an implicit calculation scheme. The nutrient cycling model uses a fully explicit scheme. In the nutrient cycling model, a time step starts with updating state variables using rates and flows from the previous time step. Next, auxiliary variables are updated and new rates and flows are calculated. The simulation starts at time=0 with state variables read form the input files.

The simulation program DYNAMITE is written in PASCAL (TURBO version 4.0, Borland Co.). The name of code of the main program DYNAMITE is DYNAMITE.PAS. In the main program the time step loop is controlled (from time=0 to time=finish time), and the calls to the PROCEDURES are made. The codes of the PROCEDURES are stored in so-called UNITS.

A UNIT in TURBO PASCAL-4.0 is almost like a separate program; it can be compiled separately and is included in the main program only when it is needed. The use of UNITS allows for the use of larger programs.

Table III.A.1 shows the UNITS with the corresponding PROCEDURES, and a short description of the function of the PROCEDURE.

The code names of the units are of the type "UNITname".PAS. After compilation, using the TURBO PASCAL-4.0 compiler, "UNITname".TPU files are produced. Some UNITS use other UNITS; e.g. the output from the water balance sub-model, WATER, is used in the growth model GROWTH.PAS. The UNIT DECLAR contains all the global declarations and is used by all other UNITS. Most variables are declared in the UNIT DECLAR.

During compilation the main program DYNAMITE.PAS scans the active DIRECTORY for all the "UNITname".TPU-files. If the UNITS are not present, they will be produced from the "UNITname".PAS-files. The UNITS are linked with DYNAMITE.PAS and written to the executable file DYNAMITE.EXE. Now the program is ready for execution, and can be run without the compiler on a IBM-compatible DOS-PC, by typing "DYNAMITE". If one wants to make changes in the programme, the TURBO PASCAL-4.0 compiler will be needed to make a new .EXE version of the program. Changes in the input files can be made with any text-editor allowing input and output of ASCII-files.

Table III.A.1 Units and procedures in DYNAMITE

Unit	Procedure	Description
DECLAR	- -	contains all global declaration
INIT	.HEAD	puts heading above output files
	.READING	reading and echoing of input file
	.INITIAL	initialization procedures
ORGDECOM	.DECCON	initialization of initial organic pool contents and decomposition/conversion rates
NWPTOTCH	.NWPOOL	calculation of new pool contents as result of all flows within a time step
	.TOTALI	adding of rates to total rates over the simulation period
	.CHECKSUM	checking of nutrient balance
RATS	.RATIOS	calculation of element ratios in organic pools
WATER	.WATERBAL	water balance calculations
DEPERWEA	.DEPOSITION	calculation deposition per time step
	.EROSION	calculation of erosion of organic and inorganic pools
	.WEATHERIN	calculation of weathering rates
DEC	.ORGDEC	calculation of decomposition and transfer from organic pools
LITTER	.LITDEC	litter decomposition and transfer + calculation of total decomposition + weathering + deposition flux to solution
GROWTH	.UPTAKE	calculation of uptake, growth, leaf fall, wood fall and root turnover
LEACH	.LEACHI	calculation of nutrient leaching flux and new solution pool content
OUT	.OUTPUT	output to output files

Running the program DYNAMITE

The simulation program is started by executing the DYNAMITE.EXE file on the MSDOS-PC.
Before starting the actual simulation the question "desired output files?" appears on screen. Typing "1", will produce 18 separate output files for various state variables, flows and ratios of N, P, K and C. Typing "2" produces 3 files in which total inflow, outflow and element content of the ecosystem during the simulation are given (checksums). An overview of the produced output files is shown in Table III.A.2.
If option "1" is chosen after finishing the simulation with DYNAMITE, the program ALLOUT can be used to group several of the output files to one or more new output files. This is done by calling ALLOUT. Several output options appear on screen and output is written to TOT[option].PAS.
The second question after starting DYNAMITE is whether initial organic pool contents and decomposition/conversion constants are input or should be calculated. When it is the first time the program is run, answer "YES". Calculated contents and constants are stored in the output file DECPAR.OUT. In a second run, if time step and total element content of the soil are the same, the answer can be "NO" and values for pool contents and rate constants will be read from DECPAR.OUT.
Now the actual simulation will start. Output to the screen and screen control is programmed in the procedure OUTPUT.

Table III.A.2 The output files of DYNAMITE

Name unit no.	Element	Compartment	Variables
PSTATE.OUT 10	P	total system	state variables
PFLSS1.OUT 11	P	soil & forest floor	rates & flows
PFLSS2.OUT 12	P	soil & forest floor	rates & flows
PFLWSV.OUT 13	P	vegetation	rates & flows
NSTATE.OUT 20	N	total system	state variables
NFLSS1.OUT 21	N	soil & forest floor	rates & flows
NFLSS2.OUT 22	N	soil & forest floor	rates & flows
NFLWSV.OUT 23	N	vegetation	rates & flows
KSTATE.OUT 50	K	total system	state variables
KFLSS1.OUT 51	K	soil & forest floor	rates & flows
KFLSS2.OUT 52	K	soil & forest floor	rates & flows
KFLWSV.OUT 53	K	vegetation	rates & flows
CSTATE.OUT 30	C	total system	state variables
CFLWS1.OUT 31	C	soil & forest floor	rates & flows
CFLWS2.OUT 32	C	soil & forest floor	rates & flows
CFLWSV.OUT 33	C	vegetation	rates & flows
CPRAT.OUT 40	C/P	soil & forest floor	auxiliary variables
CNRAT.OUT 41	C/N	soil & forest floor	auxiliary variables
PCHECK.OUT 61	P	total system	checksums
NSTATE.OUT 62	N	total system	checksums
CSTATE.OUT 63	C	total system	checksums